PRINCIPLES OF ENVIRONMENTAL SCIENCE AND ENGINEERING

P. VENUGOPALA RAO

Director
Sridevi Women's Engineering College
Hyderabad

PHI Learning Private Limited

Delhi-110092
2023

₹ 595.00

PRINCIPLES OF ENVIRONMENTAL SCIENCE AND ENGINEERING
P. Venugopala Rao

ISBN-978-81-203-2893-8 (Print Book)
ISBN-978-93-5443-053-4 (e-Book)

Published by Asoke K. Ghosh, PHI Learning Private Limited, Rimjhim House, 111, Patparganj Industrial Estate, Delhi-110092 and Printed by Syndicate Binders, A-20, Hosiery Complex, Noida, Phase-II Extension, Noida-201305 (N.C.R. Delhi).

CONTENTS

PREFACE

After independence, irrigation and industrial sectors were given priority in addition to drinking water and drainage schemes. Over the years, people realized that exploitation of surrounding environment has caused irreparable damage to ecosystems and living conditions. Pollution has become an inevitable result of progress. Mankind is facing serious environmental problems like global warming, desertification and changes in climate. Pollution Control Boards and various non-governmental organizations are taking all measures for protecting the human environment. At this stage, there is a felt need for a subject like Environmental Studies for all undergraduate students. The University Grants Commission has prescribed the essential guidelines in this regard. This textbook presents a simple and systematic approach to meet the needs of all categories of students to create total awareness on all aspects of Environmental Science and Engineering.

Chapter 1 is introductory, where biotic, and abiotic components essentially related to the sustenance of mankind in the biosphere and elements of self-regeneration capacity of the environment are briefly discussed. Chapter 2 gives a detailed account of various natural resources such as land (minerals, food, forests), water and energy and the effects due to excessive utilization of these resources for commercial gains. Chapter 3 deals with the structure and function of an ecosystem with its material and energy movements for maintaining a stable and self-sustainable environment. In Chapter 4 on Biodiversity, the value and the need for preserving the richest reservoirs of plants and animal wealth are specified against threats from the human activities. Population welfare measures and the principles of environmental sanitation are discussed in Chapter 5 on Community Health.

Chapter 6 discusses various aspects of pollution (sources and control measures) due to different impurities (physical, chemical and bacteriological) released during domestic, industrial or agricultural activities. Other problems associated with solid waste disposal, noise pollution, etc. are also presented briefly. While urbanization, industrialization and agricultural practices cause local problems, global warming and ozone layer depletion, hazardous wastes pose problems of serious environmental concern to the entire human race on the earth. These are outlined in Chapter 7. Most of the environmental problems need management solutions. Chapter 8 on Environmental Management deals management studies in industries and development projects. It also provides a note on the legal measures and activities of the Pollution Control Boards. The ultimate goal of Environmental Studies is to develop a sustainable environment, i.e. a positive relationship between the man and the biosphere. This issue, i.e. progress without pollution, is discussed in Chapter 9 on Sustainable Development.

Topics on chemistry and microbiology involved in this interdisciplinary subject are discussed in Chapters 10 and 11, whereas Chapter 12 on Monitoring and Analysis of Pollutants illustrates simple steps of analysis of various pollutants in a college laboratory. Some projects of interest and glossary have been added to benefit the students.

It may not be easy to achieve an equitable distribution of wealth or utilization of natural resources. But provision of basic facilities, like safe drinking water supply, toilets and comfortable surroundings, should be made possible at any cost to all the population. Prevention against pollution is a safe and economically-wise principle. Whether a polluter pays, beneficiaries share the cost or the government takes care is left to the rule of justice and common sense.

The author acknowledges Dr. (Smt.) A. Priyamvada Devi, Head, Department of Zoology, St. Ann's Degree College for Women, Hyderabad, for writing topics on ecosystem and biodiversity. Thanks are also due to friends Prof. C. Radhakrishna, Prof. K.V.G.K. Gokhale, Prof. K.V. Sharma, Prof. G. Jagmohan Das, Prof. G. Ram Prasad, Prof. K.M. Subrahmanyam, Sri B. Madhusudhan Rao, Sri K. Kameswara Rao for their useful discussions and encouragement. Sri K. Narsing Rao, Sri V. Satyanarayana and Sri M. Rajagopal have provided the necessary secretarial assistance.

The author wishes to express his gratitude to Sri K. Ashok Rao, Chairman, Sri Devi Women's Engineering College and to M/s Prentice-Hall of India, New Delhi for giving the author a valuable life after retirement.

The author is also indebted to his family members for providing a very happy domestic environment.

P. VENUGOPALA RAO

Part One

ECOSYSTEMS AND POPULATION WELFARE

CHAPTER 1

BIOSPHERE

TOPICS COVERED

- ❑ Constituents
- ❑ Biogeochemical Cycles
- ❑ Human Activities
- ❑ Sustainable Development

1.1 CONSTITUENTS

Water, air, food, shelter and the whole life supporting systems are drawn only from the surrounding environment. Our environment consists of land, water and a number of resources, its living and non-living components and their interactions. For the sustenance of life on the earth, Sun is the ultimate source of energy for the entire system. Life is not same at different places due to so many variations in the environmental factors. Living organisms also disturb the environmental conditions. It is a natural process where populations continuously evolve balancing their quality of life and growth. Due to the process of changes, new species may originate, some living organisms adjust with the living circumstances and only fittest may survive. Formation of earth with its various resources and the total evolution system is a slow process taking millions of years. But man can adversely change the sustainability of ecosystems and the ecological balance in a short interval of one or few decades of time itself.

Life supporting system for plants, animals and other living organisms is climate specific. Rainfall and temperature at a place depend upon factors such as latitude, altitude, nearness to the sea, intensity and duration of sunlight, topography, and wind movements. Formation of forests or deserts depends upon the amount of rainfall and the temperature at a place. Forest ecosystems are different from marine or fresh water biomes, with diverse situations in between.

Biosphere may be treated as the **atmosphere** (air—oxygen, nitrogen, carbon dioxide), **lithosphere** (land—minerals, salts, foods, nutrients) and **hydrosphere** (water—dissolved oxygen, salts) which influences and supports the entire biotic and abiotic life systems. Plants, animals and microbes utilise the nutrients consisting of C, N, S, P and other compounds which are again re-circulated due to biological and chemical reactions in nature.

Atmosphere

Dense and humid layers of atmosphere near the surface of the earth have high temperatures due to the radiations from Sun. Troposphere is the space up to a height of about 12 km from the surface of the earth where the atmospheric temperature decreases with increase in altitude at the rate of about 6.5°C per km (Figure 1.1) Air movement, cloud formation, particulate matter (dust), water vapour, gaseous pollutants and interactions with living organisms are more in this region.

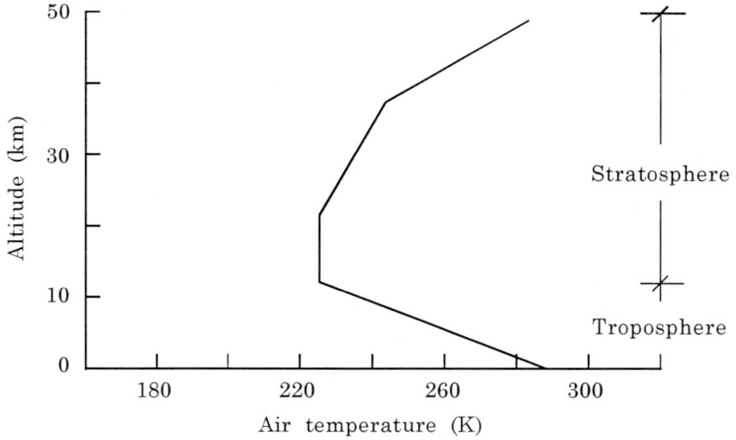

FIGURE 1.1 Air temperature vs. altitude.

Stratosphere is a 20 km zone above troposphere where the temperature increases with elevation up to a level and where ozone layer absorbs ultraviolet radiation from Sun.

Air movement is due to the following factors:

(i) Movement of earth around the Sun
(ii) Unequal temperatures in different parts of the earth's surface
(iii) Differences in the rate of cooling and heating of water and land surfaces and
(iv) Local meteorological conditions

Air moves horizontally due to planetary circulation, variations in temperature and pressure, local conditions (land-sea, mountain-valley) and seasonal wind movements [Figures 1.2(a), (b) and (c)]. A mass of air also moves in a vertical direction depending on the temperature profile, humidity

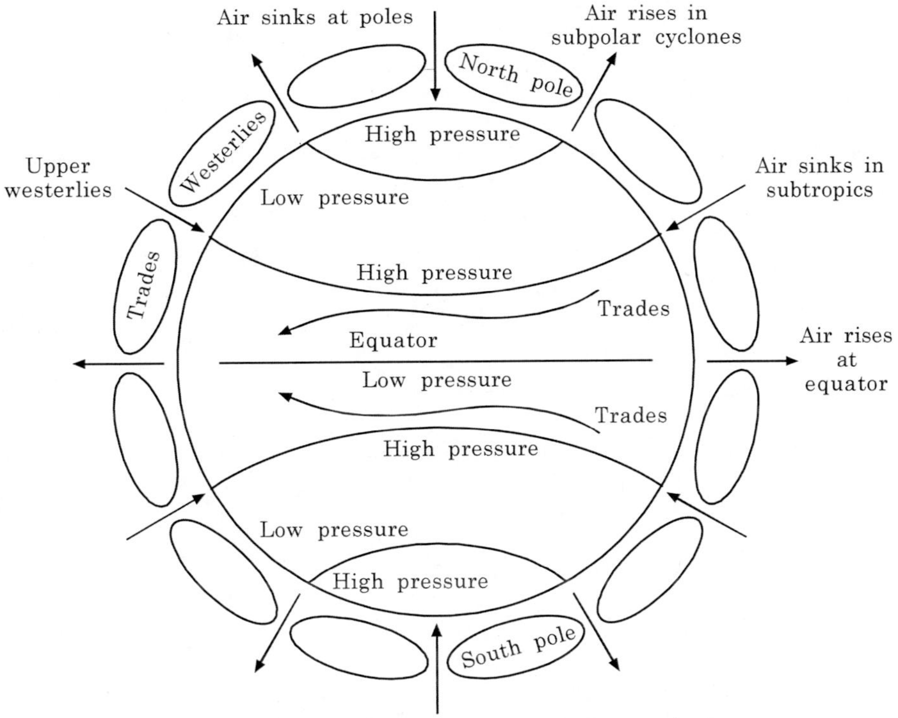

FIGURE 1.2(a) Air movements and circulations.

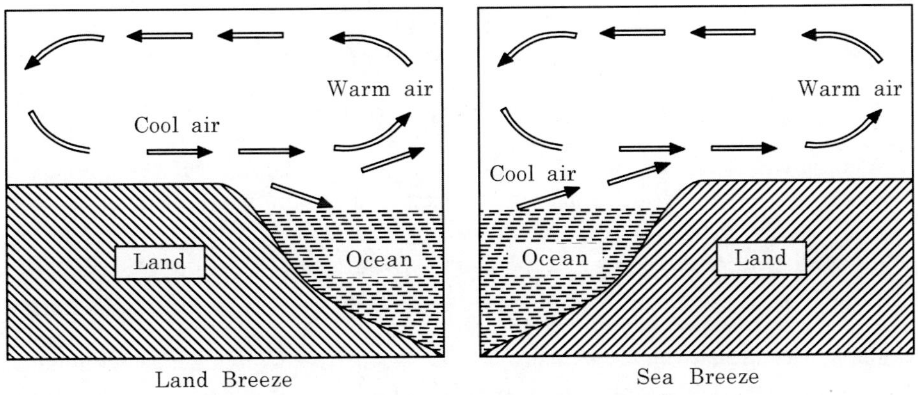

FIGURE 1.2(b) Land and sea breezes.

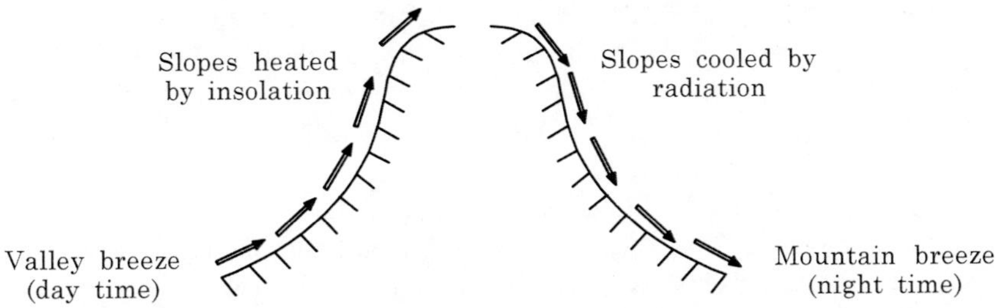

FIGURE 1.2(c) Mountain and valley breezes.

conditions and density. Air movement and mixing helps in dispersion and dilution of various particulate and gaseous pollutants. Weather conditions may vary in a certain locality on different days, but climate is likely to be location specific. Temperature, rainfall and climate influence the characteristics of life on earth. In hot and dry climates, deserts are formed due to low intensity of rainfall and higher rate of evaporation. From equator to poles, different types of forests (plant life) are formed due to variations in temperature and rainfall. Human activities such as deforestation, combustion, industrialization may have an impact on the local climate causing adverse effects on the life supporting systems.

Lithosphere

The core, crust and the top 100 km of materials of the earth (Figure 1.3) are formed over billions of years due to a number of internal and external processes

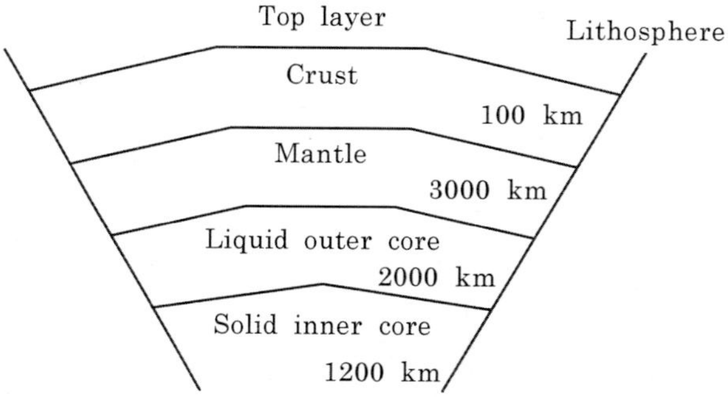

FIGURE 1.3 Structure of the earth.

of geological, chemical and biological actions in nature. Mostly the earth consists of a complex combination of metallic and non-metallic minerals such as silicates of Al, Fe and Mg, carbonates of Ca and Mg, sulphides of Fe and Pb. Temperature differences, air and water movements, oxidation or reduction, combination or decomposition, hydrolysis or salt formation, are all classified under *physio-chemical factors*. Microorganisms also interact with oxygen, sulphur, carbon and others in many ways. Ocean basins, continents, mountains, rock formations, landforms are all dynamic and continuously change with time. By providing all the essential requirements like food, shelter, water and nutrients, land has become a suitable place for the entire biota to live.

Different type of rocks exist such as igneous (basalt, granite), sedimentary (shale, limestone) and metamorphic (slate, marble), and are composed essentially of oxides and silicates. Minerals containing Fe, Zn, Cu, and Pb also occur in certain locations on the earth.

Landform includes mountains, plateaus, plains, volcanoes, ice-covered arctic and antarctic regions. Major biomes on the land are tundra (grass lands), taiga (conifers), temperate as well as tropical forests (with different trees of varying sizes), wetlands (marshy swamps), fresh water (lakes, rivers) and marine life.

More diversified plant and animal life is available in tropical forests. Water and temperature on the earth control the following:

(i) Climate
(ii) Soil characteristics
(iii) Nature and abundance of plant life
(iv) Aquatic life
(v) Human activities

The most important layer of top soil contains minerals, organic matter, inorganic substances, microbes, fungi, algae, protozoa, insects and worms. Air space in the soil contains moisture, nutrients and microorganism which decide the structure, composition, characteristics and fertility of the soil. Top soil is formed due to chemical and biological changes and weathering processes (disintegration, erosion, deposition). However the rich top soil may be carried away due to wind movement, river water flow, changes in land use patterns, urbanization, industrialization, construction and other human activities. The extent of depletion and degradation of land depends on its use.

Hydrosphere

Almost three-fourth of the earth is covered with water bodies, such as fresh water biomes (lakes, rivers), marine waters and ice. Estuaries are complex zones of river water mixing with sea. There is a transition zone between terrestrial and aquatic systems also. Oceans are a rich source of food, minerals and energy. Coastal systems are also rich in productivity.

Only one per cent of water available on the earth is fresh water which is useful to all human activities directly. Earlier civilizations developed near the water bodies, the major portion supports aquatic life. Water bodies also control the climate of a place.

Precipitation, run off, percolation, storage, evaporation, transpiration, humidity and condensation forms a water cycle (Figure 1.4) in solid, liquid and vapour stages. Major quantity of water in this natural system is available for use as small portion will be under the process of recycling.

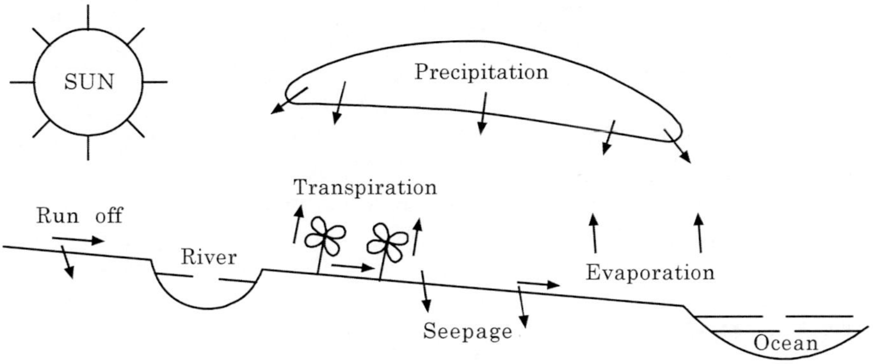

FIGURE 1.4 Water cycle.

Water bodies may be classified as rivers (lotic system) and lakes (lentic system). River water transports and disperses all materials due to continuous flow and mixing. More oxygen is available in the top layers of water. Shallow waters (littoral zone) and open waters (limnetic zone) will have rooted plants or floating autotrophs (planktons), as plenty of sunshine is available. In the bottom layer of a lake, more nutrients (decaying organic matter) may be available, but without sunshine. However, difference in temperature between upper layers (epilimnion) and lower layers (hypolimnion) may cause little vertical mixing even in stagnant lakes.

Aquatic life depends upon the following factors:

(i) Depth of water and intensity of light penetration
(ii) Dissolved oxygen and mineral salts
(iii) Temperature and solubility
(iv) Inorganic and organic chemicals
(v) Bottom deposits and water movement

Human activities cause pollution of water bodies, eutrophication of lakes, silting of reservoirs, submergence of lands, depletion of water sources and degradation of water quality.

1.2 BIOGEOCHEMICAL CYCLES

Biosphere is a bed of resources, activity, utilization and replenishment. Environmental resources are limited asset under a continuous process of material recycling. Microbes, plants and animals play their active role in keeping the essential elements under circulation. Chemical reactions are assisted, accelerated and completed by biological activity. Living systems are unique in their own way. But all the systems are interconnected in many ways in the big universe. The system is self balancing provided the rate of utilization is less than the rate of replenishment and the storage unit is undisturbed.

Carbon Cycle

Respiration and combustion processes release carbon dioxide into the atmosphere. Anaerobic decomposition results in the formation of methane. Carbon dioxide dissolves in water and combines with Na, Ca, Mg elements to form carbonates, bicarbonates and salts, They are deposited as natural minerals. Photosynthesis needs carbon dioxide for the preparation of glucose and its products. Organic matter is stored in plants, sediments, sea waters and mineral deposits. Biodegradation and plant respiration also produces carbon dioxide. More CO_2 means more photosynthesis. But CO_2 is a green house gas, absorbing infrared radiation and increasing global temperatures. Carbon dioxide is taken in and given out in a cyclic manner and forms carbon cycle (Figure 1.5).

Nitrogen Cycle

Nitrogen is essential for plant growth. Conversion of nitrogen compound is a chemical process, fully complemented by biological activity. Plenty of nitrogen

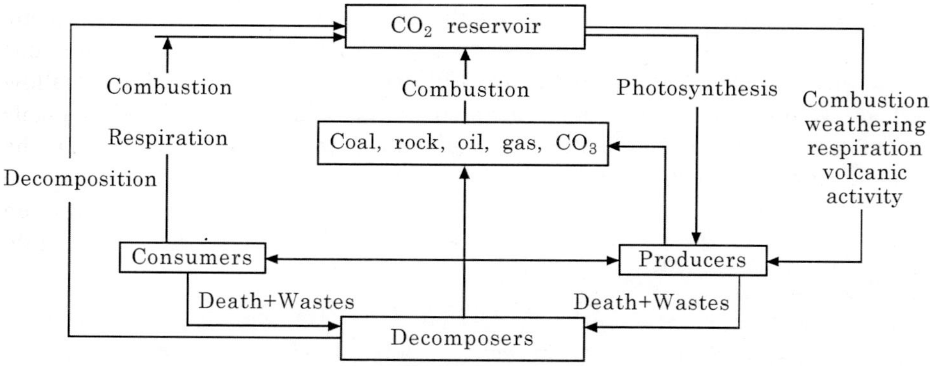

FIGURE 1.5 Carbon cycle.

is available in the atmosphere but plants cannot use directly the same. Fixation of nitrogen is an energy-requiring process. Biological fixation is done by microbes (*Azato bacter, Clostridium, Rhizobium*) or blue green algae, in soil and water.

Bacteria decomposes nitrogen in protoplasm into inorganic form. Some of this, is also converted into ammonia and nitrate which can readily be used by green plants. Nitrification is the process of conversion of ammonia into oxides of nitrogen, yielding energy. Bacteria utilizes this energy in the preparation of organic matter. Nitrification is done by Nitrosomonas and Nitrobacter. Nitrates are returned back to the atmosphere as nitrogen by *Pseudomonas* bacteria (Figure 1.6).

Nitrates in excess, contaminate drinking water sources.

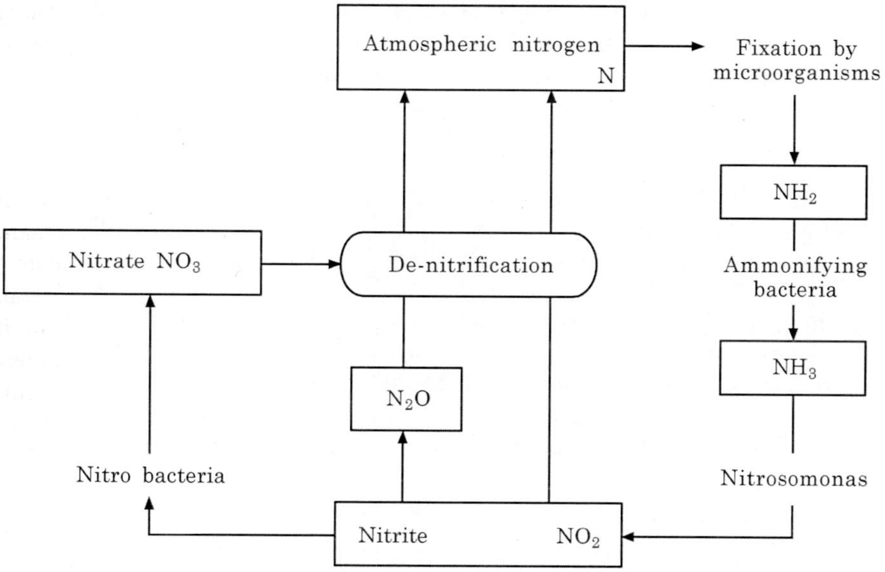

FIGURE 1.6 Nitrogen cycle.

Oxygen Cycle

Oxygen is essential for the evolution and protection of living systems. Ozone releases molecular oxygen into the atmosphere. Again oxygen molecule is converted into ozone due to the reactions with O and OH radicals. Plants also release oxygen into the atmosphere.

Phosphorous Cycle

Phosphates as mineral deposits in the ecosystem are available to autotrophs. Through vegetation phosphates reach soil. Animals obtain phosphates through grazing. Decay and mineralization releases phosphates back into the system. Excess of phosphate causes eutrophication of lakes.

Sulphur Cycle

Elemental sulphur gives sulphur dioxide and sulphates on oxidation and hydrogen sulphide on reduction. Sulphur compounds are always available in the rock formations.

1.3 HUMAN ACTIVITIES

Basic needs of food, clothing, shelter and energy requirements have to be met from the environmental resources in the ecosystem consisting of water, air and land. Man is dependent on biosphere. Human activities interfere with nature. There should be a mutually beneficial symbiosis between man and earth. A tropical forest wetland landscape has its own advantages in preventing soil erosion or floods. Environment is like a goose that lays golden egg every day. Human population should carefully use the resources and also preserve them for future generations.

Some basic facts are presented here:

(i) Resources are limited, so use them wisely and save for tomorrow.
(ii) Matter cannot be generated afresh. It should be recycled and reused to the extent possible.
(iii) Energy can be converted from one form to another but some percentage is lost into the environment.
(iv) Diverse things co-exist beautifully in an ecosystem. They are inter-dependent and maintain a self-balancing system.
(v) Never treat environment as a market, resources as commodity, population as consumer and the total system as a business. Understand that environmental resources are permanent assets that replenish and purify themselves. Hence conserve, show restraint and use the accrued interests only.
(vi) All are having equal rights and responsibilities over all resources. Avoid exploitation and misuse.

Pollution (depletion and degradation of resources) depends on the developmental activities of populations. The more the population, the more are the chances of pollution as shown in Figure 1.7.

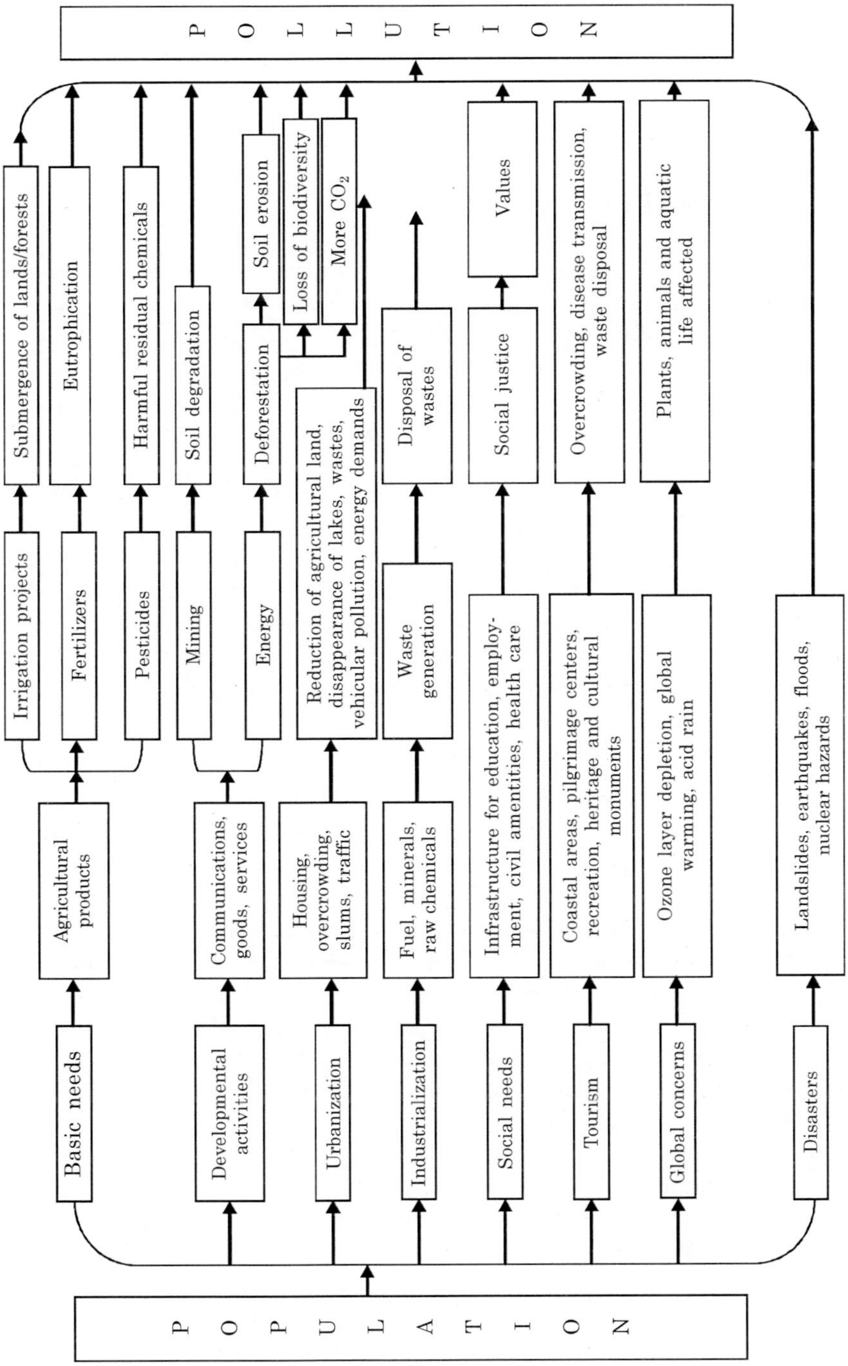

FIGURE 1.7 Population vs. pollution.

Increase in human population leads to:

(i) Increasing demand of resources
(ii) Generation of wastes
(iii) Release of harmful chemicals into the biosphere
(iv) Ecosystem disturbances
(v) Reduction in biodiversity
(vi) Depletion and degradation of resources

1.4 SUSTAINABLE DEVELOPMENT

All developmental activities are needed for the growth of an individual and the community. But today's progress should not lead to tomorrow's pollution. Quality and quantity of resources should not decrease with time. Is it possible to quantify the environmental assets and recognize the danger of degradation in terms of economic losses? Can we realize that forest, river, lake, sea are not only a source of tourism, recreation, food, fisheries, but also responsible for rains, oxygen in the atmosphere, soil fertility, energy source? Today, people are purchasing water and a day is not far away when oxygen is also to be purchased, unless the environment is kept free from pollution.

The fundamental rule is—never exceed the natural limits of carrying capacity of the environment. No doubt, basic needs are to be satisfied, developmental activities are to be taken care and standard of living should be very high. It must not cause depletion and degradation of resources. There must be principles, policies and strategies for sustainable development.

Protect the natural resources by preserving biodiversity, maintaining eco-balance and improving wild life and forests.

(a) The first and foremost need is to control population. Growing population in large numbers and accelerated growth and development are detrimental to the interests of conservation of ecosystem resources.
(b) Identify the misuse, overuse and exploitations of resources in agriculture, industrial and domestic sectors, and rectify the local conditions. Prevent leakage, wastages and use efficiently adequate quantities for genuine purpose, depending upon the availability.
(c) Reuse, recycle, replenish and conserve to the best extent.
(d) Small projects are likely to cause less damage when compared to big projects.
(e) Distribute the resources equitably to protect social justice.
(f) Afforestation, use of bio-fertilizers, drip irrigation, need based biopest management, vegetation, bunds on contours lines are some sustainable solutions in agricultural field management.
(g) Creation of a buffer zone of protection between industrial and residential zones, coastal area regulations, people and wildlife, help in protecting or separating one from the other.
(h) Protect the environment by controlling the pollution.

(i) Watershed and river quality management helps to maintain water quality and quantity.
(j) Alternate energy sources like wind power, solar heat and biomass, reduce pressure on conventional fuels.
(k) Care can be taken while selecting and using raw materials to minimise the pollution levels.
(l) Public awareness, training programmes, involvement of voluntary organizations may be useful.
(m) Legislation, rules, incentives, penalties may be adopted to sustain the environment while encouraging development.

SUMMARY

Living environment is different at different locations due to variations in latitude, altitude, nearness to sea, sunlight (intensity and duration), topography, rainfall, temperature and climatic conditions. But both biotic and abiotic systems support and maintain a healthy relationship for the sustenance and continuance of the whole life on earth. That is why the essential elements like carbon and nitrogen are recycled in nature, and water and air are self purifying to a greater extent. However some human activities intended for progress and development are exceeding and disturbing the carrying capacity of the environment. It is time for developing sustainability for the welfare of mankind.

KEYWORDS

Atmosphere, biogeochemical cycles, carrying capacity, hydrosphere, lithosphere, pollution, progress, sustainability of the environment.

SOLVED QUESTIONS

1. Write a note on earth's evolution.

Three thousand million years ago, only anaerobic organisms lived due to lack of sunlight, blanket of CO_2 and absence of O_2. Formation of amino-acids led to primitive life as large mass of hot earth got divided into continents and seas in course of time. Hydrological cycle is responsible for shaping the land scape, weathering of rocks, transportation and deposition of sediments. Hot core of earth created convection currents leading to continental drift. Original land mass split into two major continental groups, isolating India and Australia; Africa and South America. India moved northwards resulting in the formation of Himalayas. Natural events like earthquakes, volcano's or Tsunami waves may change the entire world map. Many physical, chemical and biological processes are established. Lands and rocks (minerals) are formed. Due to weathering process, small particles slowly matured into top fertile soil.

2. **Explain the meaning of 'ecobalance'.**

   Plants and animals utilize oxygen from the atmosphere for respiration, liberating carbon dioxide. Photosynthesis is a food making process during which plants utilize CO_2. Ozone dissociates into molecular oxygen in the atmosphere. Combustion releases CO_2. Forest trees, sea water and carbonate compounds are sinks for CO_2. Thus the ecosystem maintains an O_2 and CO_2 (oxygen: carbon dioxide) balance.

3. **Give two examples of human interference with nature.**

   (a) Houses are constructed on the beds of lakes reducing the area of water bodies, ultimately leading to disappearance of lakes.
   (b) Illegal activities (hunting, smuggling) are causing loss of forest cover and biodiversity.

4. **Indicate two cases of pollution in a town.**

   (a) Disposal of domestic wastes without adequate treatment
   (b) Smoke and gaseous emissions due to vehicular traffic

5. **How are afforestation programmes useful in sustainable development?**

   Forests are mainly responsible for the utilization of a green house gas CO_2 in the atmosphere and conservation of biodiversity. Forest products are a source of support for different living systems on the earth. Deforestation leads to soil erosion, floods, silting of water bodies, drought and change in climatic conditions. Curb the exploitation of forest resources to sustain the system.

6. **State two movements in which the public have participated in the conservation of environment.**

   (a) *Chipko movement* in which trees were protected by the villagers
   (b) *Narmada movement* in which submergence of lands are being prevented

7. **Comment on the statement, 'material recycles in nature'.**

   Living systems use the food that is available in nature for its metabolic activities. Biotic systems decompose and join the nature again after death. Inorganic substances (minerals, metals etc.) are also renewed in the environmental system. The whole process is recycling nutrients even though some energy is lost in the mass transfer systems.

8. **What do you understand by El Nino phenomenon?**

   Shift in atmospheric pressure over the pacific ocean resulted in the reversal of trade winds affecting rainfall and weather conditions. This led to disruption of normal patterns causing floods or droughts at different locations.

REVIEW QUESTIONS

1. Write a detailed note on the factors affecting climate at a place.
2. Describe the effects of nature on mankind.
3. List out the factors that affect life supporting system on the earth.
4. Mention the various human activities that cause degradation of natural resources.
5. What are the steps taken for maintaining sustainability of water resources.

OBJECTIVE TYPE QUESTIONS

1. The part of the earth and its atmosphere in which organisms live is
   (a) Biosphere (b) Biome
   (c) Biomass (d) Biota
2. The part of atmosphere between stratosphere and thermosphere is
   (a) Mesosphere (b) Ionosphere
   (c) Exosphere (d) None
3. Tundra climate is related to the region
   (a) Arctic ocean (b) South East Africa
   (c) Mediterranean sea (d) Northern Australia
4. Atmosphere consists of 79 per cent N_2 and 21 per cent O_2 by
   (a) Volume (b) Weight
   (c) Density (d) All the three
5. CO_2 is essential to living systems (Yes/No).

CHAPTER 2

NATURAL RESOURCES

TOPICS COVERED

- Introduction
- Land Resources
- Mineral Resources
- Food Resources
- Forest Resources
- Water Resources
- Energy Resources

2.1 INTRODUCTION

All living beings need air, water, food, shelter, clean surroundings and other basic life supporting requirements for carrying out physiological, socio-economic, cultural and other activities. Only the environment can supply the necessary requirements in varying forms at all times. Land is the proper place for all elements to maintain and sustain a living relationship in the universe. Coal, oil and a number of useful minerals are present in the earth's bed. Plants and animals on the land or in the water bodies provide food to human beings. Forest, a place for wildlife, not only contributes to the commercial activity, but keeps the ecosystem in balance.

Resources are essential for the development of life on the earth. They should be properly utilized for a comfortable living in the society. The entire living system is likely to collapse when the consumption exceeds the regeneration capacity, or if the resources are not uniformly distributed at all places and to all sections of population or if the resources are very limited in quantity or when the resources are polluted. Commercialization, over-exploitation, expansion of agricultural operations, industrialization, use of coal, petroleum products, metals and various chemicals to satisfy the need and greed of increasing populations are the major reasons for the rapid depletion and degradation of all natural resources, which are formed over a very long period of time.

Resources can be classified as:

(i) *Biotic*: plants, animals, livestock, fisheries, etc. giving food, milk, meat and other products for the sustenance of human life.
(ii) *Abiotic:* fossil fuels, metals, mineral deposits, water, land and other elements which support living system.

Some resources are *renewable* i.e. replenished within a reasonable time or capable to be present for a long time. Solar energy, atmospheric nitrogen, oceans, wind movement, biomass are some examples. Water, land and biota become non-renewable in some cases depending upon exploitation or more use at a rapid rate.

Some resources are *non-renewable* i.e. cannot be replaced when once they are exhausted. They are available in limited quantities. It is not easy to increase the quantity or to regenerate in a reasonable time interval. Fossil fuels, minerals and metals from geological origin come under this category.

Top soil and river water may be degraded from quality considerations due to human activity. Nuclear energy or tidal power is not affected by human activities. Gold and platinum may be reused but lead (in petrol) or zinc (in galvanizing process) cannot be recycled for use. Continuous harvesting is possible if soil is fertile but sometimes some species are endangered. Population is a resource but some countries are over populated. Genes are preserved to create new breeds of plants and animals and protect biodiversity. Waste also is a resource when it can be purified, recycled and reused in some form or the other. Most of the human beings are responsible for the quantity depletion and quality degradation of natural resources. Increasing populations demand more and more resources for their basic needs. New scientific and technological advances help in the effective utilization of resources to improve the living standards. Resources are meant for use only. Commercial exploitation cannot be prevented easily. However if utilization exceeds the sustainable limits of growth, the ecosystem is likely to collapse. So resources should be utilized with care and should be preserved for future. Conservation does not mean 'stop the use', but prevent misuse and abuse of resources as they are limited and permanent assets for human population. Right to use does not give right for polluting and lowering the quality of life on the earth.

Some attempts to protect the resources are as follows:

(i) To reduce demands
(ii) To control population growth rate
(iii) To put resources for optimum needs
(iv) To reduce waste products
(v) To recover or reuse with proper care
(vi) To develop technologies for cost effective alternate resources

2.2 LAND RESOURCES

Land provides the required rich nutrient medium for the growth of plant life. Top soil is involved in agricultural operations as it contain humus. Water is stored in ground layers. A number of metallic, non-metallic substances are found as deposits in the land. Land is a habitat for living beings. Forest products have commercial value and also provides food. In the following sections, soil formation, land degradation and waste land reclamation studies are in detail. Mineral, forest and food resources are also discussed in the subsequent sections.

Soil Formation

Soil is the upper fertile layer of the earth that can be ploughed for the growth of plants. It is the basic raw material formed by the weathering of rocks, through various actions in nature. *Weathering* is the process of disintegration and decomposition of rocks and minerals at or near the surface of the earth through physical (mechanical) or chemical actions into smaller particles. The factors of physical weathering may be due to atmospheric changes in temperature and pressure, abrasion, erosion or transportation of finer material and spreading of plant roots into soils. There is no change in the chemical composition on the soil formed from the parent rock. Coarse grain soils such as gravel and sand are formed by the process of physical disintegration. They are called *cohesionless soils.* In chemical weathering process, atmospheric gases and moisture, carbon dioxide, nitrogen, hydrogen, oxygen in the presence of moisture react with the surface of rock. The resulting product has a different chemical composition and a poor physical constitution as compared to the original rock. The chemical weathering process involves one or more processes such as solution, hydration and hydrolysis, oxidation and reduction, base exchange formation of colloids and carbonation. Final products are clay minerals which impart cohesive nature to soils.

Soils formed at a place may be transported to other place by agents such as water, wind, gravity and other forces. Soils carried and deposited elsewhere by water are known as *alluvial sediments*, e.g. Delta lands. Soils which are carried and deposited by wind are known as *acolian deposits*. Sand dune is an example. Soil material is further subjected to climatic and atmospheric conditions. As rainfall begins to leach through the soil material for giving a habitat to small organisms, vegetation starts to develop. Gradually with the passage of time, intensive formation takes place in the characteristics of soil. Thus soil profile is a natural succession of strata below ground surface extending to different depth at different places. Thickness of each stratum varies. The top horizons (up to 50 cm) are rich in humus and original plant residue which are more valuable for agriculture. Soil fertility depends upon its structure, organic content, moisture content, designated by the engineering properties such as porosity and permeability.

Porosity is the property of retaining moisture content in the pores of the soil and permeability is the property of allowing water to flow through the soil pores.

Soil Erosion

Soil erosion is the removal of top soil by the transport agents like wind and water. When rain falls on a gentle sloping ground, loose soil particles float and move along with flowing water. It results into uniform skimming of top layer. This continuous erosion is not uniform at all places. So patches or gullyings occur on the land surface along the flow line. Loss of top soil and terrain deformation is mainly due to water movements. Soil erosion occurs when vegetation that binds the soil is removed. In dry regions like Rajasthan, sand storm remove top layers of soil. Landslides occur on mountain slopes due to natural forces like water and wind, causing soil erosion.

Land Degradation

A change in the characteristics which affect fertility of soil is termed as *degradation*. Soil erosion is a major problem. Water logging and salinity render soil conditions unfit for raising crops due to less drainage facilities. Deforestation, overgrazing, agriculture practices and industrialization contribute to soil degradation at a place.

The area under cultivation may be reduced when agricultural fields are used for other purposes such as housing projects and laying roads. Crops may not give sufficient yield if the top soil contains chemical residual due to excessive use of fertilisers and pesticides. Deforestation is taking place at a faster rate due to increasing demands for timber, fuel and forest product because of urbanization and industrialization.

Land degradation means:

(i) loss of nutrients from fertile soil
(ii) less vegetation cover
(iii) change in soil characteristics
(iv) pollution of water sources
(v) ground water contamination
(vi) changes in climate

Prevention and Control

(i) Agricultural Practices

(a) Crop rotation, in which different crops are grown in the same area following a rotation system i.e. for some time rice and later other crops are grown which helps in replenishment and better nutrient utilization.

(b) In strip farming, cultivated crops are sown in alternate strips (to prevent water movement) during the period in the same field.
(c) Ridge and furrow type of irrigation lessens water run off and soil erosion.
(d) Cultivation of grass land improves the soil structure within the field crops.
(e) Mulches (protective layer of plants) keep the soil moisture fixed, preventing evaporation loss.

Afforestation is the best solution to prevent soil erosion. Ploughing aerates the soil pore.

(ii) Engineering Practices

(a) Trenches or ditches are excavated at suitable location on a sloping ground to divert excess water from the affected areas.
(b) Terrace bunds (or contour farming) across the slope of the hill side are useful in collecting and diverting the run off to avoid erosion.
(c) Small check dams are constructed using stones and other materials to reduce the velocity of run off so that soil supports vegetation.
(d) Water flow may be restricted along channel by constructing suitable bunds or banks.

Waste Land Reclamation

Land sickness is due to various reasons as explained earlier. Waste lands are unproductive and causes irreversible damage. Stress due to population demands or pollution of environment may be reduced to some extent possible. However strategies for waste land development may be outlined as follows:

(i) Encourage social forestry (with babul, arecanut, amla, neem, imli) but not for commercial activity.
(ii) Use less chemicals in agricultural operations.
(iii) Create vegetative hedges and construct water harvesting structures for sustainability.
(iv) Give rest to the land for some time without crops and allow it to recover mineral loss.
(v) Practice bioremediation to restore the land.

2.3 MINERAL RESOURCES

Minerals are formed in the earth through geobiochemical processes in a long span of time. There is no chance of early replenishment. India is self sufficient in coal, bauxite, iron and manganese. The reserves are as follows:

Coal	–	200	billion tonnes
Bauxite	–	3	billion tonnes

Iron	–	12.0	billion tonnes
Copper	–	7.0	billion tonnes
Gold	–	0.2	billion tonnes
Lead	–	2.0	billion tonnes
Zinc	–	2.0	billion tonnes

Other minerals available are: Chromium, Silver, Tungsten, Asbestos, Barytes, Gypsum, Granite, Limestone, Mica, etc. Metallic and non-metallic minerals are distributed at different locations in states of Bihar, Andhra Pradesh, West Bengal, Rajasthan, Gujarat, Karnataka, Madhya Pradesh and Orissa (Figure 2.1)

FIGURE 2.1 Mineral Wealth of India.

As population increases and demands are pressing, mineral consumption increases rapidly. In a span of 30–50 years, depletion of mineral resources is likely to take place in our country. Non-renewable resources are to be conserved only. Three options for conservation are as follows:

(i) Use less and prevent wastage.
(ii) Recycle and use.
(iii) Choose alternatives.

The use and throw system exhaust the mineral resources. Then the problems of trade, foreign exchange, political pressures and economics play a major role.

Mining consists of various operations like location of deposits, extraction, purifying, refining, processing into useful products and transportation. Any activity involved in mining is likely to cause the following effects on the environment:

(i) Disturbing landscape, forests and wildlife
(ii) Releasing toxic pollutants
(iii) Water and air pollution
(iv) Land degradation

During and after mining, environmental problems are to be solved immediately.

2.4 FOOD RESOURCES

Human beings require food to grow, reproduce and maintain good health. Without food, bodies cannot stay warm, built or repair tissue or maintain heart beat. Eating right food (carbohydrates, proteins, fats, minerals, vitamins) helps to avoid certain diseases or recover faster when illness occurs. 5 per cent reduction of fluids in human body affects muscle and brain formation. Herbivores live on plants, grains (rice, wheat, corn) and agricultural products. Carnivores eat meat, fish, poultry products in addition to vegetables and grains. Though sufficient quantity of food is available, hunger, malnutrition and disease are common in developing countries as poor people do not have the purchasing capacity. Some times, the subsidized food through fair price shops may not reach all consumers. With increasing population, per capita food availability will be less. **Green revolution** started with high yield, fast growing crop varieties, putting more land into cultivation, using more of pesticides and fertilizers, construction of irrigation projects etc. But probably population is increasing at a higher rate than the agricultural production with all the development projects. There will be always an upper limit for the environmental inputs.

Fertilizers (N & P) join water sources causing eutrophication, increasing aquatic plant growth. Pesticides (chlorinated hydrocarbons, organophosphates, carbamates, pyrethrum, methyl bromide, carbon tetrachloride, etc.) enter food and water sources. They reach human beings magnified through the media (water, crops or fish products) and cause harmful effects. Food poisoning may also be due to *Staphylococous* and *Botulinum* bacteria.

Oceans also provide food, minerals, energy and number of related products besides navigation facilities.

Some solutions for food crisis are as follows:

(i) To increase land area for cultivation
(ii) To control population growth rate
(iii) To improve soil fertility and use biofertilizers
(iv) To prevent soil erosion, water logging and soil salinity
(v) To develop forests and wildlife
(vi) To give subsidies to small farmers
(vii) To maintain sustainability in the environment

Livestock and Fisheries

Cows, buffaloes, goats, sheep, horses, camels, donkeys, pigs and a number of animals and birds also live in the surrounding domestic environment. Meat, milk, eggs are useful food components. Animals are useful as alternate systems of transport at some places. They also help in contributing the organic manure.

Fish (including crabs, prawns, shrimps) from oceans, rivers, lakes or water tanks or developed in aquaculture ponds are also a source of food supply to the human beings. As the Indian coastal line is vast, about 10 million tonnes of fish are likely to be available yearly.

2.5 FOREST RESOURCES

About 20 per cent of land area is covered with forests, that is around 60 million hectares in India and 6 million hectares in Andhra Pradesh. It comes to 0.08 hectare per capita in India whereas the world's average is 0.64 hectare per head. Temperature and rainfall (moisture content) are two important parameters which decide the nature of forests at a place. (Table 2.1)

Table 2.1 Different types of forests

	Forests	*Moisture content*	*Temperature*	
(i)	Tundra	Low moisture	Low temperature	Arctic
(ii)	Deciduous and Coniferous (Taiga)	High moisture content	Low temperature	55°– 65° latitudes
(iii)	Tropical or seasonal	Heavy rainfall	Moderate temperature	30°–30° latitudes
(iv)	Grass lands of Savannas	Moderate or less rainfall	High temperature	Equator
(v)	Desert	No moisture	High temperature	20°–30° latitudes

Dense and diverse forests grow near the equatorial latitudes. In other zones, the nature of forest depends upon the extent of rainfall, soil fertility, climate and seasons. Conifers and temperate forests are seen in North America and Eurasia. Tropical rain forests are available at some parts of Central America,

Australia, Indonesia and temperate shrubs at mediterian area. Tropical seasonal forests are grown at places in South central parts of America, South eastern parts of Asia, India and deserts in North Africa (Sahara) and Mexico.

These forests grow at different places on different latitudes and altitudes (Figure 2.2) from equator to poles. In tropical coastal areas, near estuaries where salt water or fresh water is in abundance and nutrients are available in plenty, most productive mangrove forests grow. Forests are responsible for maintaining the ecological balance and biogeochemical cycles in nature.

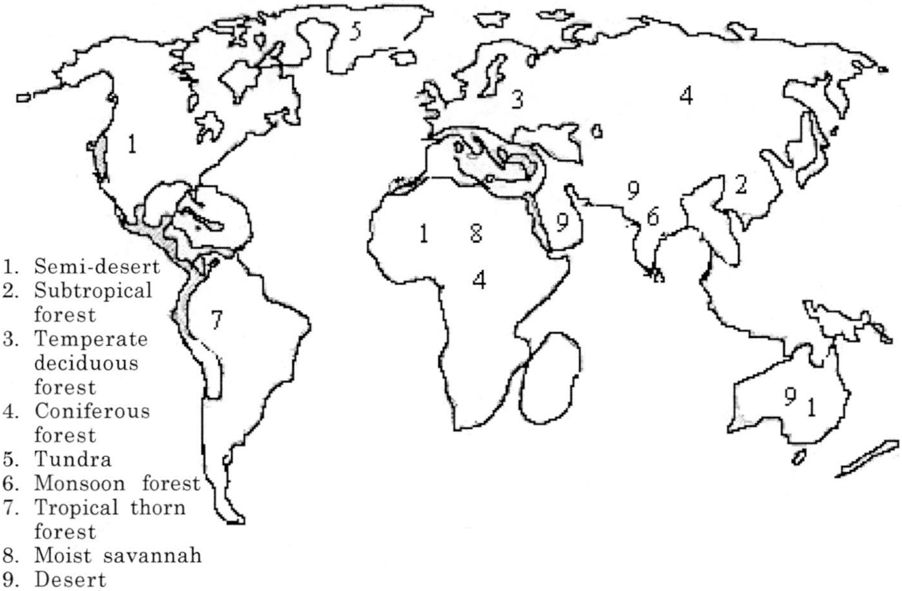

FIGURE 2.2 Distribution of forests.

Functions of a forest are listed below:

(i) Soil erosion, floods and droughts can be prevented,
(ii) CO_2 is utilized and O_2 is released into the atmosphere,
(iii) Heat balance on earth is maintained by absorbing solar radiation and greenhouse gases,
(iv) Coastline is protected,
(v) Pulp, resin, fibre, gum, honey, medicine and other valuable products are obtained,
(vi) Forests are the main source of timber and fuel wood,
(vii) They provide the habitat for wild life and preserve biodiversity,
(viii) They are the source for knowledge, recreation and tourism, and
(ix) Forests are developed for economical gains.

Inspite of a number of advantages, 2 per cent forest area is removed every year. Some causes of deforestation are as follows:

(i) Increase in population numbers, their needs and life styles

(ii) Urbanization and industrialization
(iii) Mineral exploration
(iv) Construction of dams and reservoirs which submerge forest lands
(v) Human encroachment for timber use, agricultural operations and illegal trade activities
(vi) Overgrazing

Demerits of Deforestation

(i) It leads to change in the climate. As rainfall reduces and temperature increases, desertification starts.
(ii) It leads to loss of food products, useful material and livelihood.
(iii) It causes loss of biodiversity and imbalance in ecosystem.
(iv) It leads to global warming.

Hence, there is a pressing need for protecting and improving the forest lands for the overall benefits of humanity. People should understand cutting one tree (more than 10 years old) will never be equal to planting 50 seeds now.

Wild Life Protection Act (1972) and *Forest Act* (1980) positively help to preserve forests and wild life. One should balance short term and long term gains and the regeneration capacity of new plants and the removal rate of aged plants. Social forestry or afforestation programmes are involved in the intensive plantation schemes for sustainability. The desirable goals are to prevent commercial exploitation, to grow different categories of plants and to conserve plant growth in parks and sanctuaries, to reclaim waste lands and monitor all forest welfare programmes.

2.6 WATER RESOURCES

Hydrological Cycle

Water is the most essential commodity for all living beings for their survival and development. It is required for various domestic and industrial needs, irrigation and power supply, navigation, recreation and aquatic life. Uneven distribution (place and time) of rainfall leads to floods and droughts at some places. Increase in population and their standards of living are mainly responsible for depletion and pollution of water bodies. Serious problems are occurring due to the non availability of adequate supply of quality water sources in the country. Conservation, purification and recycling are the main options for obtaining continuous supply of water.

Hydrology is concerned with the transportation of water limited to the release of water from clouds or precipitation and re-entrainment of water into clouds through evaporation from different water surfaces (Figure 2.3). About 20 per cent of rain falls on the land surface and flows as run off, as the major portion of the earth is covered by sea. Out of this, only 22 per cent reaches the

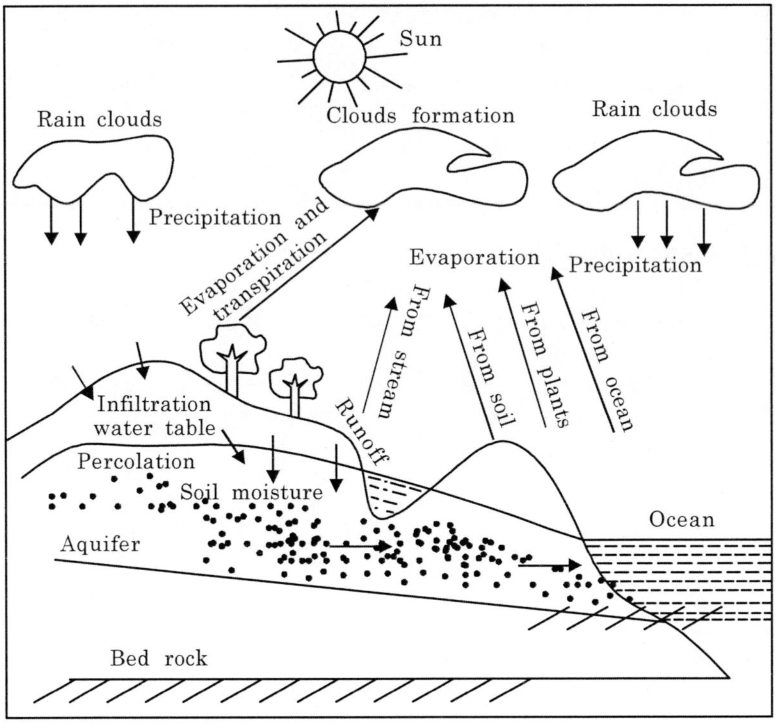

FIGURE 2.3 Hydrological cycle.

surface sources such as rivers, lakes; 10 per cent percolates into the ground; 25 per cent is retained as soil moisture in the top soil and the rest is evapotranspired from land, water and other surfaces. Surface run off depends upon the characteristics of the catchment area, such as topography, ground slope, soil characteristics, ground water table, land use patterns. A substantial portion of the rain water is getting infiltrated into the ground and gets further percolated into deeper layers for storage. Movement of ground water is very slow. Water that is held in the top layers of the soil under capillary forces is known as *subsurface water* which is very important for crops.

Out of 170 million hectare metres of annual surface flow, 80 per cent portion joins the sea. Total annual discharge in major rivers like Brahmaputra or Ganges exceeds 5×10^6 cubic metres of flow and Mahanadi, Godavari or Krishna may have an annual run off of less than 1×10^6 cubic metres.

In India, length of important rivers and main canals is around 10 lakh kilometres. Lakes and resources are about 3 million hectares of surface area. The ground water storage appears to be to an extent of 40 million hectare metres, but only a small fraction of water is available for use.

Aquifers are water bearing strata below ground level. Water can be pumped from shallow or deep wells. Heavy discharges from deep ground layers raise in pumping costs, change the soil characteristics, affect the climate, increase the dissolved salt contents (like fluorides) in water and may also cause favourable condition for landslides or earthquakes.

Diversion of water from lakes and silting of reservoirs may also alter the landscape.

Water is used for irrigation, domestic, industrial needs and also to generate electricity. Irrigation needs 5,000 to 15,000 m^3/hectare depending upon the crop variety and domestic consumption is 150–300 litres per capita per day based on the standards of living. Major portion of water (about 70 per cent) is used for irrigation purpose, 15 per cent is used for power generation and only 5 per cent is consumed in the domestic sector. Some parts of Punjab and Tamilnadu utilize more than 75 per cent of ground water sources. All these numerical values of rainfall, run off or ground water resources are average figures only.

People experience floods at some places during monsoon months whereas some are continuously drought prone areas. In reality, availability of water is location specific and sometimes seasonal. However the total quantity of water present in the environment is a finite and fixed quantity. Only a small fraction of that quantity is available for utilization in different sectors. Similarly, small percentage of available water resources takes part in the hydrological cycle. The major share of water storage that cannot be directly utilized, also contributes to the ecological balance on the earth and so is an asset. Another significant point of interest is that water flows do not observe physical or geographical boundaries such as neighbouring districts, states or countries. Floods or pollution at one place may affect people over one hundred kilometres on the downstream side.

River Valley Projects

These are the projects on the valley portion of river flows essentially meant for storage of water during flow season (monsoons) or floods and supply the all time needs to meet the agricultural demands for raising different types of crops through a network of canals. They are also useful for generating electricity, supplying drinking water, raising fish culture, navigation and recreational purposes.

The component structures of a river valley project as shown in Figure 2.4 are as follows:

(i) Dam—to stop and contain the flowing water as a reservoir
(ii) Supply sluice—to regulate the supply of stored water to the different needs through canal network
(iii) Spillways—to safely dispose the flood water or excess flows into the river so that no damage is done to these structures
(iv) Canals—to convey water to far off places for meeting the water requirements

The highest water level up to which water can be stored is known as full reservoir level. Water coming into the reservoir above this level has to be disposed off with the help of a surplus weir or spillway. Water stored above the sill level of the supply sluice is released into the canals.

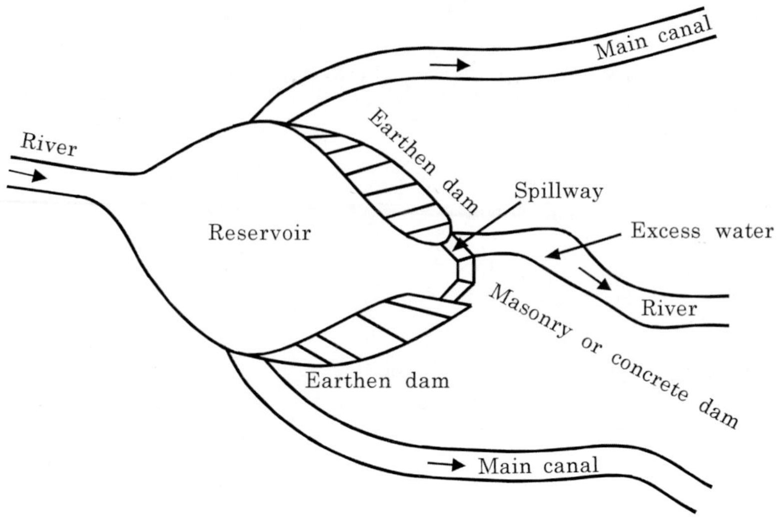

FIGURE 2.4 River valley project.

Water storage improves greenery and also changes the climatic conditions of the surrounding environment. Waterlogging and salinity affects soil fertility. Water storage leads to mosquito breeding. Large dams are supposed to create earthquakes in the nearby regions and submerge more areas.

Floods and Droughts

(i) Due to intense rainfall or melting of snow, rivers get filled up with water and overflow breaches canals or tanks, submerged lands and destroys life and property. Floods also cause soil erosion and change the land use patterns. Flood flows can be estimated, flood timings can be predicted and loss of life can be prevented to some extent by adopting water resource management systems. Forests regulate water run off, reduce soil erosion, retain soil moisture and thus help in controlling floods.

(ii) Drought is lack or insufficiency of rain (less than 400 mm/year) for an extended period that causes water shortage, depletion of ground water levels and soil moisture, reduction in stream flows and creates hydrological imbalances. It is the most serious hazard to agricultural production in all parts of the world. It leads to serious economic consequences and untold human misery. Water is not available even to satisfy basic human needs in such areas. Some areas are highly drought prone in the country.

Cloud seeding for rains is quite expensive and unpredictable. The better alternative appears to be to conserve resources, use carefully and recycle water after proper treatment. Another effective method is to resort to watershed management practices.

Priorities and policies may be framed to link river waters depending upon the feasibility and conserve ground water adequately so that both flood and drought situations are prevented.

Watershed Management Practices

Watershed is the area, draining the precipitation and a synonym for a catchment of a river basin. The important characteristics of watershed are the size, shape, type of drainage and the distribution of rainfall. Some other considerations are degree of ground slope, land use pattern, vegetation, geology of rock strata and soil characteristics and meteorological factors.

Watershed management is a concept of developing a technology to use every drop of rain water at the place of occurrence. It can be location specific as per the needs like social forestry, soil conservation, crop management or ground water storage. An integrated approach for the watershed management should include protecting natural resources for attaining the desired results.

The following steps are involved:

(i) Data collection with details of hydrology, drainage and topographical features of an area over a good length of time

(ii) Data analysis and planning for the measures to be adopted for:

  (a) Soil conservation and green cover

  (b) Ground water recharge such as construction of check dams, percolation tanks and rain water harvesting measures

  (c) Improvement of drainage and utilization of surface storage

(iii) Working out and tallying the water balance in the watershed by measuring evapotranspiration, precipitation, run off, recharge to ground water, valley storage, irrigation, ground water draft, infiltration and out flow

(iv) Implementation of the schemes by construction of structures, encouraging social forestry, artificial recharge, spacing of wells and drip or sprinkler system of irrigation

Deforestation should be avoided. Open drains are to be improved in plains. Conjunctive use of surface and ground water prevents water logging. Proper drainage and ground water pumping is necessary to preserve water quality and maintaining crop yield.

Water or moisture content should be preserved in top soil to keep the nutrients safely in porous soils. Soil conservation and vegetation cover, improves the ability of land to hold water. 10 mm rainfall in a small farmer's land about a hectare, amounts to 100,000 litres of water, if stored through water harvesting programmes.

Soil conservation measures include contour bunding, ploughing, (at right angles to hill slope), fallowing (no crops), trenching, terracing (series of steps) and soil reclamation. The objectives of water harvesting are to conserve rain water, to recharge ground water source and to increase agricultural productivity.

On sloping grounds or plains, vegetation reduces run off. Contour bunds, trenches, vegetative barriers prevent land slides on mountain slopes. Desilting of existing tanks on a regular basis is the simple and best method of improving water storage capacity.

Water harvesting recharge pits can be constructed in court yards in individual houses to collect rain water from roof tops. Small bunds may be constructed to prevent the movement of rain water to the adjoining fields.

While considering the water conservation measures in a particular area, it is necessary to make liberal provision for the downstream interests and the riparian rights. If a number of watersheds on the upstream side adopt conservation, storage and recharge measures are taken, the irrigation or water supply tanks or storage structures on the downstream will suffer. It is not possible or advisable to utilize the entire water of a river basin without serving the interests of the downstream areas. Is it possible to control floods without releasing water from river to downstream regions? That is why water resources belong to the entire humanity. Ecological crisis can be prevented by the justified equitable distribution of natural resources in any democratic set up.

2.7 ENERGY RESOURCES

The actual energy consumption in the five sectors of energy (domestic, industry, agriculture, transport and commercial) is estimated to be around 380 kg of oil equivalent per capita per year in India. Traditional biofuels (fuel wood, crop residue and animal waste) contribute 30 to 40 per cent of the total energy consumption, in the form of non-conventional energy. The rest of the energy is supplied as conventional energy in the form of coal, oil, gas and hydro and nuclear based electricity. Figure 2.5 shows energy consumption in the form of coal.

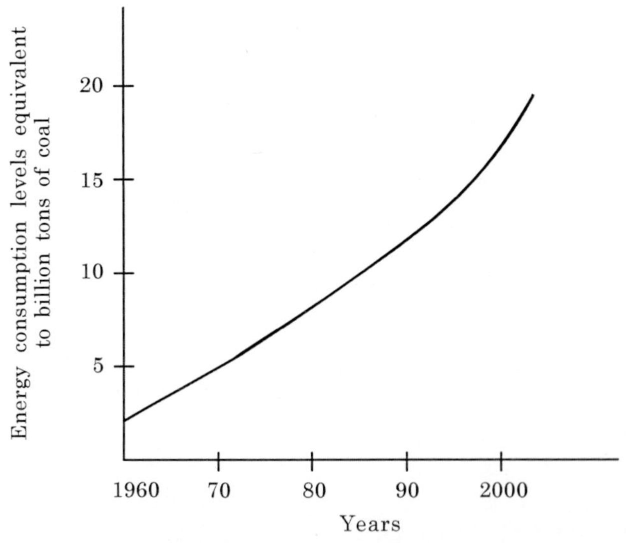

FIGURE 2.5 Energy consumption.

Fossil fuels are formed from decomposed plant and animal matter found in various strata below earth's surface as carbon bearing (hydrocarbon) coal, oil and natural gas. Estimating these energy supplies require geological knowledge, economical calculation and technical know-how. With varying degrees of uncertainty, current annual production levels imply that there are nearly 200 years of coal, 65 years of natural gas and 40 years of oil at the present production and utilization levels. These fossil fuel resources are not evenly distributed around the world. Nearly 70 per cent of the world's proved recoverable oil resources are in the Middle East, 4.3 per cent of natural gas reserves are in the Commonwealth of Independent States and 25 per cent in the Middle East. Two thirds of hard coal reserves are in North America and CIS. In India, coal is available in Andhra Pradesh, Jharkhand, Chattisgarh, Orissa and West Bengal and gas reserves are available in Assam, Gujarat and Andhra Pradesh. Presently, India has an installed power capacity of about 110,000 MW of which 20,000 MW is based on diesel oil. Coal and lignite contribute about 61,000 MW and 25,000 MW are from large hydroelectric power projects. Production is less than consumption needs. When compared to the high rate of consumption levels, the rate of regeneration of resources is very low. It is estimated that in India, per capita electricity consumption is about 400 units (KWH), while the world's average is about 1600 units and in advanced countries it is more than 10,000 units.

Energy resources in India are estimated as follows:

(i) Coal and lignite—200 billion tonnes (approximately 25 per cent of the total resources used)
(ii) Crude oil—800 million tonnes (30 per cent)
(iii) Natural gas—700 billion cubic metres (20 per cent)
(iv) Nuclear production—400,000 tonnes
(v) Hydro electric power—90,000 MW (8 per cent)
(vi) Biomass—300 million tonnes/year (10 per cent) (renewable)
(vii) Wind—20,000 MW (renewable)
(viii) Ocean—80,000 MW (renewable)
(ix) Solar energy—terrestrial flux of 500 W/m^2

Oil, natural gas and coal can be considered non-renewable as the fossil fuels are replaced in nature at a very slow rate when compared to the human developmental needs. In fact, the radioactive fuels Uranium and Thorium are also transformed over a very long time span of billions of years. Renewable energy resources are the solar energy, wind, ocean, geo-thermal heat and vegetable or animal matter. When the energy consumption is increasing at a very rapid rate, fossil fuel resources are depleting accordingly. This alarming difference in demand and supply rates has raised serious energy problems.

Fossil Fuels

These resources are concentrated in a relatively few regions of the world. They are non-renewable (exhausting) sources. Oil price and supply depends upon import policies and pressures from various quarters. Transportation also is a major factor for consideration.

Coal based thermal power plants use pulverized firing technology with an overall efficiency less than 60 per cent. Indian coal (30–40 per cent ash content, 1 per cent sulphur and other impurities) generates 60 million tonnes of fly ash annually. About 1–2 per cent is used for briquetting and the rest is disposed off in landfills. The calorific value is 5000 kilo calories per kg of coal, combustion efficiency is low. Gaseous (NO_x, SO_x) and particulate emissions into the atmosphere also cause considerable pollution. CO_2 is a greenhouse gas and SO_2 causes acid rain.

Hydel Power Plants

Large dams are constructed on major rivers for utilizing the water for irrigation and generating electricity. Nagarjunasagar, Bhakra Nangal, Hirakud, Tungabhadra, Damodar, Mettur and Koyna are some projects. However some disadvantages like water logging, salinity of lands, siltation of reservoirs, submergence of forest lands and villages, displacement and rehabilitation of people affected and possibility of earthquakes have been experienced. A number of differences of opinion amongst different parts of the people and controversies over Narmada, Tehri, Almatti and other projects have given preference to small dams only. Power generated by these projects is also less than 8 per cent and river water is not available in plenty to generate the turbines for the production of sufficient quantum of power throughout the year.

Traditional fuel wood collection and its use and large hydroelectric schemes are widely regarded as unsustainable. Modern utilization of bio-mass (organic matter) involves large scale removal of trees (eucalyptus) or the conversion of crops into fuels (sugarcane into ethanol), may adversely affect biodiversity, natural habitats and the quality of the environment. This has led to the utilization of non-conventional and renewable sources of energy. They are also clean to a reasonable extent.

Wind power developments can be bulky and obtrusive. Wind speed i.e. kinetic energy (mechanical power) is converted into electrical power with about 25 per cent to 30 per cent efficiency. Wind power has emerged as successful renewable technology with about 1200 MW installed production in some locations of Gujarat and Tamilnadu. Geothermal energy project (transforms natural heat in the crust of the earth) may release CO_2, H_2S and Hg. North-western Himalayas are the sites with hot spring water. Tidal power projects may destroy or fundamentally change estuarine habitats. Biogas based power plants are also installed at some places. A potentially important additional source of energy is nuclear energy. More than 30 countries have 430 commercial nuclear power plants in operation in France and other locations, with a total generating capacity of about 340 giga-watts.

Sun is the major ultimate source for solar energy, biomass and fossil fuels. The total solar energy resource is over 10,000 times the world's current annual primary energy use. The key question is: how much of this alternate non-conventional, renewable energy will be realized in the coming decades?

Biogas

Microbial activity releases methane by decomposing organic matter under anaerobic (without air) conditions. Biogas microbes consist of a large group of complex and differently acting microbe species, notably the methane producing bacteria. Biogas contains 65 per cent methane and 30 per cent carbon dioxide and can be used for cooking (heating) and lighting purposes. Wood, agricultural residues, sugarcane, garbage and animal dung may be the inputs for generating biogas.

Process

In the first step (hydrolysis), bacteria decomposes the long chains of the complex carbohydrates, proteins and lipids into monosaccharides, peptides and amino acids. In the second step, acid producing bacteria, convert the intermediates of fermenting bacteria into acetic acid, hydrogen and carbon dioxide. These bacteria are facultatively anaerobic and can grow under acidic conditions. To produce acetic acid, they need oxygen and carbon. For this, bacteria utilize oxygen in solution form, further creating anaerobic conditions essential for methane production. Moreover, compounds with low molecular weight are reduced into alcohols, organic acids, aminoacids, carbon dioxide, hydrogen sulphide. This process is **endergonic** (i.e. only possible with energy input) from a chemical reaction considerations. Finally methane producing bacteria decompose the compounds to methane and carbon dioxide. These three stages of *acid production*, *acid regression*, and *alkaline fermentation* are aided by suitable micro organisms at different steps.

Biogas plants (in a small scale) are adopted in small sector households using agricultural residues and animal wastes. The gas is utilized for cooking and domestic lighting purposes. Municipal garbage (biomass) has been converted into generating electrical power in city units and towns with reasonable success to supplement power needs. In Andhra Pradesh, there are 37 biomass power plants generating 1200 million units of power annually.

Nuclear Energy

Matter can be changed into energy. Heat energy is released (which is converted into electricity) by splitting (fission) the nucleus of an atom. Two hydrogen atoms can be combined (fusion) to form one helium atom to release energy. These actions are chain reactions that continue further and further.

Uranium and Thorium are used for nuclear power generation. Plutonium is used in fast breeding reactors. Kalpakkam (Tamil Nadu), Tarapur (Maharashtra), BARC (Trombay), Narora (UP), Kaiga (Kerala), Manuguru (AP) are some of the locations connected with nuclear power plants. There is presently no basic shortage of radioactive minerals for nuclear power generation. Reprocessing and recycling Uranium and Plutonium, would increase the number of years for which fuel would be available. The use of breeder reactors would greatly extend availability.

Nuclear plant construction has been installed in the place far from the human population and is concerned with operational safety, waste disposal and proliferation of nuclear weapons. Accidental leakages and failures of safety measures in nuclear power plants causes disaster in the human environment. However, nuclear energy is pollution free, and probably technically feasible for a fair trial in the future. It is renewable, eco-friendly and economical.

Ocean Energy

Oceans cover about 70 per cent of the earth's surface. This makes them the world's largest solar energy collection and energy storage system on an average day, 60 million square kilometres of tropical sea absorbs an amount of solar radiation equal in heat content to about 250 billion barrels of oil. Even if 0.1 per cent of this stored solar energy could be converted into electric power, it would meet all the electricity demands of the world on any given day. The world's ocean may eventually provide us with energy to power all our installations in future. There are three basic ways to tap the ocean for its energy, (i) waves, (ii) high and low tides, and (iii) temperature differences in water.

(i) Kinetic energy of the moving waves in the ocean, can be used to power a turbine. A rising wave forces the air out of a chamber which in turn spins a turbine connected to a generator. Up and down motions of the wave may be converted to the movements of a piston that is linked to a generator. Most of these wave energy systems are very small, but can be used to power a small light house.

(ii) Tides are the result of the gravitational force of the moon and the sun as well as the revolution of the earth. Certain coastal regions experience higher tides as a result of geographical features such as bays and inlets. A difference of atleast 5 m height between high and low tides is required for generating power. The higher the tides, the more electricity can be generated from a given site at a lower cost of production. There are about 40 sites around the world with this magnitude of tidal range.

Tidal energy is an essentially renewable resource that has none of the typical environmental impacts of other traditional sources of electricity, such as fossil fuels or nuclear power. Changing the tidal flow in a coastal region may result in a wide variety of impacts on aquatic life.

(iii) Sunlight warms ocean water. It is warmer on the surface but below the surface, it is very cold. Power plants can be built by using this difference in temperature to make energy. A difference of 25°C to 30°C is needed between top and bottom layer temperatures to create ocean thermal energy. There are some demonstration projects in Japan and Hawaii.

Solar Energy

The total amount of solar energy is responsible to create all other types of resources like biomass, ocean, wind, hydropower systems. The heat energy is stored and utilized. Solar cookers, water heaters or desalination units have glass plates or panels to directly collect and transfer heat. Active solar systems need mechanical power to circulate water or air from collectors. Passive systems without mechanical devices are used for energy needs in the domestic sector.

Photovoltaic cells convert solar radiation into electrical power that can be stored or transferred. These cells are used for lighting systems, pumping water and small power plants. In rural sector, solar energy finds a number of uses. It can be effectively used in hostels, hotels and public institutions. Probably a solution for water shortage appears to be the installation of desalination plants in coastal areas. Solar energy is non-polluting, but the intensity of heat is not uniformly available at all places and all times.

Fuel Cells

Fuel cells are energy conversion devices that theoretically has the capability of producing electricity as long as the fuel and oxidant are supplied to the electrodes. Gaseous fuels are fed continuously to the anode compartment and an oxidant (oxygen from air) is continuously fed to the cathode compartment. Electrochemical reactions take place at the electrodes to produce an electric current. These fuel cells are known as *alkaline fuel cells, direct method fuel cells, molten carbonate fuel cells, phosphoric acid fuel cells, proton exchange membrane fuel cells* or *regenerative fuel cells* depending upon the process and materials in use.

Energy Conservation and Sustainability

The following points are to be considered while dealing with energy utilization problems:

(i) Technological feasibility
(ii) Operational efficiency
(iii) Losses in the system
(iv) Cost of generation, storage and distribution
(v) Depletion or degradation of resources or renewability
(vi) Eco-friendly or pollution free nature

Our future energy requirements will not be met by any single energy resource or unique processing technology. Renewable energy technologies play a complimentary role at least for the next 50 years. A judicious mix of fossil, nuclear and renewable resources with appropriate technologies, backed by a scientific analysis for efficiency, emissions, economy and reliability as applicable to different power capacity slots, must be developed to facilitate the optimum use of the available energy resources to meet the requirements. Energy is utilized in the domestic, industrial and transportation sectors. Energy consumption depends upon the standards of living in rural and urban areas.

Energy conservation is a management technique to identify the proper energy source, development of alternate energy scheme and select suitable energy methodology through cost benefit analysis. The objective of energy conservation is to achieve maximum energy utilization at minimum cost. Conservation does not mean a sacrifice in quality but in using energy more efficiently. It also does not mean not using energy at all.

Ideally domestic buildings should be located where topography and vegetation provide the most suitable microclimate. Building orientation taking advantage of and compensating for solar radiation and wind should be of prime importance in the construction of buildings. Also the materials used for construction should admit and intercept daylight and solar radiation that is essentially required. It should also permit exchange of air and moisture with surroundings. A building is to be designed for passive solar technology in conjunction with natural ventilation, orientation and window management, appropriately selected building materials and effective control of thermal comfort inside the building.

Coal and lignite will continue to contribute a major share of our energy consumption. Clean coal technologies are to be adopted for higher efficiencies and better environmental performance. Imported oil is expensive and a great drain on foreign exchange reserves. Nuclear power development, wind based, biomass based power generation, mini hydel plants, solar photovoltaic cell technology appears to be the future hope for the developing countries like India. Geothermal energy has a base in the Himalayan hilly areas of Ladakh. Biogas has its roots in rural development. Energy plantations (Jatropa, Pongamia) for extraction of bio-fuel constitute a long term possibility. Heat and power cogeneration technology is also a means for the conservation of energy.

SUMMARY

Biotic (plants, livestock, fisheries) and abiotic (fuel, minerals, metals) resources are meant for proper utilization but not for abuse and exploitation. Fertile top soil formation takes considerable years of time even under favourable circumstances. But excessive use of chemicals (fertilizers and pesticides), modern methods of cultivation and indiscriminate disposal of liquid and solid wastes are degrading the land productivity. Commercial exploitation of forest products and mining projects are exceeding the regeneration capacity of nature. River valley projects which are meant for a good purpose, appears to be under scrutiny. Time has come for conserving every drop of water. Energy needs are growing whereas coal, oil and gas reserves are lowering. A number of alternate energy (nuclear, biomass, wind) resources are being searched depending upon the geographical locations. No doubt, population (explosion) is also considered as a resource but that consumes all other resources. Let the day be green tomorrow also.

KEYWORDS

Energy sources, crisis and conservation, hydrological cycle, land resources (minerals, food, forest) degradation and reclamation, river valley projects, watershed management.

SOLVED QUESTIONS

1. What are the methods available for minimizing water logging conditions?

Water logging is a condition of top soil remaining saturated with water and presence of little of soil air. These lands accumulate salts and cannot support plant growth. The menace can be minimized by the following ways:

(i) Providing proper drainage to the water logged areas through desilting and resectioning of existing natural drains and through creating new drains
(ii) Lining of canals in order to minimize seepage
(iii) Pumping of water from the waterlogged areas back into the canals
(iv) Providing proper drainage facilities for the collection and disposal of irrigated waters
(v) Using ground water for irrigation in the stretches prone to water logging

2. What is the future for nuclear energy?

Atomic Energy Commission is planning:

(i) 12 numbers of pressurized heavy water reactors using natural Uranium as fuel
(ii) Construction of 4 numbers of prototype fast breeder reactors at Kalpakkam for the generation of 500 MW each, using Plutonium oxide as fuel and liquid Sodium as coolant
(iii) Advanced reactors to generate electricity using Thorium as fuel

Nuclear power is zero emission energy

3. What are the demerits of a huge multipurpose project?

A multipurpose (power, irrigation, water supply, pisciculture and tourism) project has the following impacts on the environment:

(i) Loss of livelihood, displacement and rehabilitation of inhabitants
(ii) Loss of land (agricultural, forest, monuments and valuable)
(iii) Soil erosion in the catchment area and siltation in the reservoirs
(iv) Water logging in the cultivating lands
(v) Deforestation
(vi) Induced earthquake activity
(vii) Loss of (some species) flora and fauna

4. **Explain the term 'Bio-remediation'.**

   Bio-remediation is a method of detoxifying the soil contaminants using appropriate microorganisms. Bio-remediation is to stimulate microorganisms to rapidly degrade hazardous chemical contaminants through optimization of environmental conditions.

5. **Give an account of livestock as renewable resource.**

   Cattle, goats, buffaloes, sheep, horses, camels, fowls, ducks, poultry are very important resources providing milk, eggs, meat, leather, dung and wool. They are useful in transportation, agriculture and domestic sectors. They contribute to national economy. Fisheries, birds, reptiles and a number of plants are also major sources of food supply in addition to the livestock.

6. **What are wetlands/mangroves?**

   Wetlands (saturated with water) are very fertile and productive to support all vegetations and variety of species. Mangroves are wet lands (dense forests) near coastal areas providing a natural habitat. They prevent soil erosion, protect from floods and cyclonic winds and provide essentials for livelihood.

7. **How does a nuclear reactor work?**

   When a U-235 nucleus splits into neutrons and fission products, a large amount of energy is released from a very small quantity of uranium fuel as mass is converted into energy from the reactor.

   In case of accidents (nuclear holocaust) or leakages, life cannot exist due to enormous heat generation, radiation causes genetic effects and damages to vegetations.

8. **Why are alternate energy sources not preferred?**

   The following are some reasons:

   - Available at some locations only
   - Available in limited quantities only
   - Production and running costs
   - Difficulties in storage and transportation
   - Unreliability and unknown risks

REVIEW QUESTIONS

1. Write short notes on pollution-free energy sources.
2. Write a note on some measures to prevent land degradation.
3. What are the problems associated with mineral exploitation?
4. Mention few adverse effects of green revolution.
5. Outline the measures to conserve energy consumption.

6. How are floods prevented or controlled?
7. Locate the places in India where minerals like gold, limestone, coal and bauxite are available.
8. With a neat sketch, explain the hydrological cycle.
9. Identify some areas (in the world) where Tundra, Coniferous, Savanna and deserts are located.

OBJECTIVE TYPE QUESTIONS

Fill in the blanks with suitable words:

1. Renewable energy resources account for a ________ percentage of the total power generation in India.
2. ____________ is an example of renewable energy.
3. Solar energy is stored in ____________
4. Desalinization projects are based on ____________ energy use.
5. In agricultural operations ______ portion of soil is utilized.
6. Addition of fertilizing elements to water bodies cause _________
7. Disintegration of rocks is known as ____________ process.
8. Soil transportation is the process of
   (a) erosion (b) deposition
   (c) both (d) none
9. Water logging results in
   (a) fertile soil (b) more greenery
   (c) less dissolved salts (d) none of the above.

CHAPTER 3

ECOSYSTEMS

Contributed by Dr. (Smt.) A. PRIYAMVADA DEVI,
St. Ann's Degree College for Women, Hyderabad

TOPICS COVERED

- Introduction
- Structure of an Ecosystem
- Functions of an Ecosystem
- Ecological Succession
- Types of Ecosystems

3.1 INTRODUCTION

Ecological science is concerned with the study of living organisms such as plants, animals and microbes (presence, growth and welfare) in relationship with the environment and interactions among themselves in a structural and functional unit known as **ecosystem**. Different species use the environment in different ways and coexist by mutual adjustment and development. Some interactions may be detrimental to their interests whereas the fittest survives.

The Earth consists of different regions as mountains, deserts, plains, seas, grasslands, forests, crop lands, rivers, lakes, oceans, etc. Each region constitutes an ecosystem and is physically different from the other and is inhabited by different organisms that are well adapted to their environment. An ecosystem is a collection of living organisms and their non-living environment, which interact with each other to form a natural unit of functional stability. Thus, an ecosystem is the total of community of interacting populations of different species of plants and animals and their habitat. **Habitat** refers to the physical and chemical factors of the place where the organisms live. In an ecosystem, plants and animals form the biotic (living) community and habitat forms the abiotic (non-living) environment. The size of an ecosystem is variable. It may be small as a pond or large as an ocean or a forest.

The interactions among the components of the ecosystem include the flow of energy and cycling of material. The energy from the sun is fixed by plants, which is used for the formation of complex organic compounds, the nutrients. These compounds are transferred to animals, both plant eaters and animal eaters. After the death of the plants and animals, the components of their bodies are released by decomposition in to the soil and water. The products of decomposition are taken by plants for producing nutrients again.

The different types of ecosystems are separated from each other only by space but are functionally interdependent. Energy and nutrients of one ecosystem find their way into another and keep the entire biosphere functioning.

Characteristics of an Ecosystem

1. Ecosystem is the major structural and functional unit of ecology. (Ecology is the study of organisms in relation to the environment).
2. The structure of an ecosystem is related to the diversity of its species (occurrence of a variety of species). Complex ecosystems have high species diversity.
3. The function of an ecosystem is related to the flow of energy among its biotic components and cycling of material within and outside the system. Material or nutrients are continuously exchanged among the various components of an ecosystem. The flow of energy is unidirectional and non-cyclic but the flow of nutrients is cyclic, as shown in Figure 3.1.

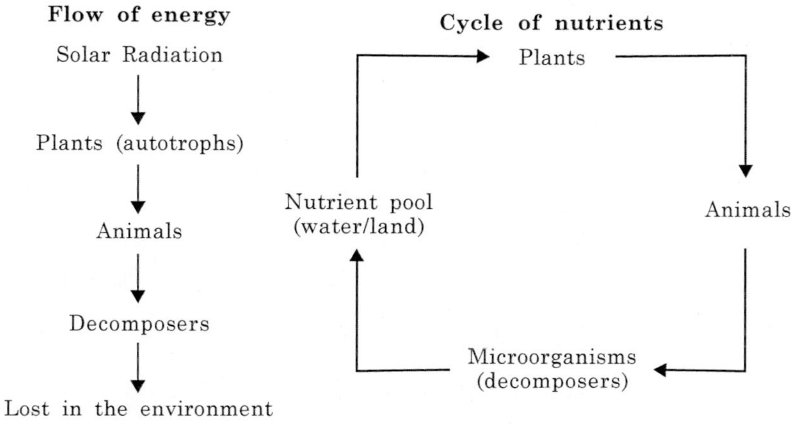

FIGURE 3.1 Flow of energy and nutrients.

4. Each ecosystem has its own energy budget. The relative amount of energy needed to maintain an ecosystem depends on its structure. Complex ecosystems, with species diversity, need less energy to maintain themselves.

5. Ecosystems maintain a homeostatic state. Certain negative feedback mechanisms help in this process. For example, when nutrients are released in excess into an ecosystem, certain chemical equilibria operate and inhibit their further release into the system.

3.2 STRUCTURE OF AN ECOSYSTEM

The structure of any ecosystem is composed of two chief components namely biotic and abiotic.

Biotic Components

The biotic components include living organisms i.e. plants, animals, bacteria, viruses, etc. The living organisms depend on the abiotic components for their survival.

The biotic components of an ecosystem are of three types according to the specific roles they play in operating the system. They are producers, consumers and decomposers.

(a) Producers or Autotrophs

Producers are the autotrophic organisms which can manufacture their own food material. They are the only organisms that can take energy from the non-living environment and also make it available to all other living organisms which cannot prepare their own food materials. They can capture solar energy to manufacture food from simple inorganic substances like water, carbon dioxide, salts, etc. in the presence of a green pigment chlorophyll. This process of producing organic food material (carbohydrates) from inorganic substances in the presence of sunlight and chlorophyll is called *photosynthesis.*

$$6H_2O + 6CO_2 \xrightarrow[\text{Chlorophyll}]{\text{Sunlight}} C_6H_{12}O_6 + 6O_2$$

A portion of the food thus synthesized is used by the producers for their growth and survival whereas the remaining is stored for future use. Photosynthetic activity contributes to food and biomass production.

Producers are of the following types:

(i) *Green plants.* e.g. *Chlorella, Spirogyra* (algae), *Azadirachta indica* (neem tree)

(ii) *Photosynthetic bacteria.* e.g. *Chromatium, Chlorobium.*

(iii) *Chemosynthetic bacteria.* e.g. *Nitrosomonas, Nitrobacter, Beggiotoa.*

The chief producers vary from one type of ecosystem to another. Some of them are given in Table 3.1.

Table 3.1 Chief producers of different ecosystem

S. No.	*Ecosystem*	*Chief producers*
1.	Ocean	Phytoplankton. e.g. Diatoms (small microscopic plants)
2.	Lake and Pond	Rooted and floating plants and phytoplankton
3.	Grassland	Grasses
4.	Forest	Trees

(b) Consumers or Heterotrophs

Consumers are heterotrophic organisms, (nourished by others) primarily animals which consume the producers directly or indirectly.

Consumers are of various types depending on the nature of the food they consume:

(i) *Primary consumers/herbivores:* Primary consumers are the group of organisms that consume the producers directly. In aquatic ecosystems, the level of primary consumption is very high, almost cent per cent. The primary consumers of different ecosystems are given in Table 3.2.

Table 3.2 Primary consumers of different ecosystems

S. No.	*Ecosystem*	*Primary consumers (Herbivores)*
1.	Most aquatic ecosystems	Protozoans, crustaceans, molluscs, etc. which feed on algae
2.	Grassland ecosystem	Rat, insect, cattle which feed on grasses
3.	Forest ecosystem	Rabbit, deer, elephant which feed on plants

(ii) *Secondary consumers/carnivores (flesh eaters):* Secondary consumers are those animals that feed upon the primary consumers. They are also called primary carnivores, e.g. dog, cat, wolf, etc. Omnivores which feed on both flesh and plants are also included in this category.

(iii) *Secondary carnivores/tertiary consumers:* Tertiary consumers are those carnivores which feed upon the secondary consumers, they are also called secondary carnivores, e.g. lions which feed upon cats and dogs, etc.

(c) Decomposers/Microconsumers/Reducers

The decomposers are also heterotrophic organisms which are saprotrophs. They consume the food by absorption but not by ingestion. They are mainly fungi, bacteria and certain protozoans. They secrete certain enzymes which can break down the complex organic substances into simple absorbable forms. This process is called *decomposition*. When the producers and consumers die, their bodies are acted upon by these enzymes and the resultant products are absorbed by the decomposers. Some of the absorbed products of decomposition are used for their own growth. The other product like inorganic nutrients, minerals and gases like ammonia are released back into the environment from where they are recycled i.e. they are again made available to the autotrophs for the synthesis of food. The continuous functioning of an ecosystem depends on the activity of decomposers in the recycling of matter.

Examples of decomposers are bacteria (*Bacillus, Pseudomonas, Clostridium*) and fungi, (*Aspergillus, Fusarium, Alternaria*) moulds and mushrooms.

On the basis of the above description, it is clear that plants are predominant producers, animals are the consumers and microorganisms are the decomposers.

Abiotic Components

The Abiotic components of an ecosystem include the non-living constituents of the environment i.e. the habitat. A habitat is a specific set of physical and chemical conditions (space, substratum and climate) that surrounds a single species, a group of species or a large community.

Soil and water form the media as well as important abiotic factors in an ecosystem. In addition to this, there are other abiotic factors which can be broadly divided into two categories.

(a) Physical Factors

Some of the important physical factors are as follows:

(i) *Light:* The chief natural source of light is the solar radiation and it provides the necessary energy to run numerous ecosystems of the earth. The quality and intensity of light varies in the major habitats depending on the cover, transparency and other factors. Green plants, through the basic process of photosynthesis, capture light energy and convert it into chemical potential energy which is passed through food chain.

(ii) *Temperature:* Temperature is the measure of intensity of heat. It controls the climate of a place in co-ordination with evaporation and precipitation. Consequently it has a direct control on the distribution of organisms.

(iii) *Evaporation and precipitation:* Evaporation and precipitation along with temperature are the main parameters, which control the climate of any geographical area and drive the water cycle. They modulate some features as salinity, hydrodynamics and biogeochemical cycles. In terrestrial environment, they modulate the development of biomes.

(iv) *Gravity:* Gravity controls the rock material cascade system and the hydrological cascade system and determines the movement of matter to and from the system. It shows distinct effect on the structure, general orientation and distribution of animals.

(v) *Pressure:* Atmospheric pressure decreases with altitude by a factor of 10 for every 15 km above the sea-level, and increases by

1 atmosphere for every 10 m depth in water. Animals have to undergo the required adaptations to this change in the pressure. Thus pressure limits the distribution of organism at high altitudes and at different depths in the sea.

(vi) *Humidity:* Humidity can be defined as the moisture content of the atmosphere. It controls evaporation and precipitation and is influenced by temperature and wind. Transpiration by plants and water absorption by animals is greatly influenced by atmospheric humidity.

(vii) *Air and water currents:* Currents of air and water are direct expressions of pressure changes. In response to this parameter, organisms develop several adaptations. For example, crabs in the sea beaches emerge with receding tide for feeding and return to their burrows as the waves sweep over the shore. Air currents play an important role in weathering process of rocks which is linked with cycling of nutrients. Air and water currents are involved in the process of 'overturn' of water which helps in rejuvenating deeper waters with oxygen.

(b) Chemical Factors

Some of the important chemical factors are as follows:

(i) *Oxygen:* Oxygen is considered to be the pulse of the environment. It is being continuously used in respiration of organisms in an ecosystem and replenished through the process of photosynthesis by producers.

(ii) *Carbon dioxide:* Carbon dioxide is the raw material for photosynthesis and its balance in nature is maintained through respiration. It controls other chemical factors such as pH, carbonates and bicarbonates of the medium in which the organisms live.

(iii) *Minerals (nutrients):* Nutrients are the elements that are essential for the maintainance of life. *Macronutrients* are those which are needed in large quantities e.g. carbon, hydrogen, nitrogen, sulphur, phosphorus, calcium, potassium and sodium. They occur in the form of simple compounds such as carbon dioxide, water and nitrates in nature. On the other hand, certain nutrients, though essential, are required in small quantities and they are called *micronutrients* e.g. iron, manganese, magnesium, zinc, cobalt and molybdenum. The source of these nutrients in terrestrial ecosystem is soil and in the aquatic system, is water. The nutrients get into the biological systems through absorption by plants or by fixation through soil microorganisms.

(iv) *Organic matter:* Carbohydrates, proteins and lipids constitute the living organisms and are also dispersed widely in the non-living form, in the environment. As the organisms die, their remains decay and become organic detritus which is taken by the decomposers that help in the recycling of elements.

3.3 FUNCTIONS OF AN ECOSYSTEM

All ecosystems maintain themselves in characteristic dynamic state. This is achieved by the production and flow of energy through their biotic components and circulation of materials like N, C, H_2O, S and P within as well as outside the system.

Thus, the functions of an ecosystem include the following:

(i) Energy
(ii) Primary production
(iii) Secondary production
(iv) Food chain
(v) Food web
(vi) Energy flow
(vii) Ecological pyramids
(vii) Biogeochemical cycles

Energy

Energy is the ability to do work. The main source of energy for an ecosystem is the radiant energy or light energy derived from the sun. The amount of solar radiation reaching the surface of the earth is 2 cal/sq. cm/min. It is more or less constant and is called *solar constant* or *solar flux*. About 95 to 99 per cent of the energy is lost by reflection. Plants utilize only 0.02 per cent of the energy that reaches the earth. Plants convert light energy into chemical energy (sugar, $C_6H_{12}O_6$) by a process called *photosynthesis*. The sugar, thus formed, is utilized for many purposes.

(a) It can be converted into starch which is stored.
(b) It combines with other sugars to form cellulose.
(c) It combines with inorganic substances as N, P, S to form amino acids, proteins, nucleic acids, pigments, hormones etc.
(d) Some amount of sugar is oxidized during respiration and energy is released to perform various functions.

$$C_6H_{12}O_6 + 6O_2 \rightarrow 6CO_2 + 6H_2O + \text{Energy}$$

Primary Production

Primary production refers to the amount and rate of energy produced by autotrophs in an ecosystem. It is expressed in terms of grams or kilocalories per square meter per day or per year. The total amount of sugar and other organic materials produced in plants through photosynthesis per unit area per unit time is called *gross primary production.*

During photosynthesis, respiration is also carried out by the plants simultaneously. During respiration some amount of sugar is utilized. Hence the net primary production can be calculated as follows:

$$\text{Net Primary Production} = \text{Gross Primary Production} - \text{Respiration}$$

Measurement of Primary Production

The primary production can be measured by one of the following methods:

(a) By harvesting the plants grown on a particular area and expressing their wet or dry weight per unit area per unit time (i.e. biomass).
(b) By measuring the amount of carbon dioxide entering and leaving the system since it gives an idea of the relation between the rate of photosynthesis and respiration.
(c) By estimating the amount of oxygen produced per unit area per unit time since it forms an index to measure the rate of photosynthesis.

Secondary Production

The energy trapped by the producers (primary production) is utilized by the consumers. The producers are directly consumed by the herbivores that are eaten by the primary carnivores that in turn are consumed by the secondary carnivores. The consumers store some amount of energy in their tissues. This energy, stored by the consumers, is called secondary production. Only about 10 to 20 per cent of the primary production is converted into secondary production. The remaining 80 to 90 per cent is lost by the consumers in the form of faeces. Thus, the rate of energy storage at consumer levels is referred to as "secondary productivity". There are three fundamental aspects of secondary productivity: standing crop, material removed and production rate.

(a) Standing crop: The amount of living material in different trophic levels or in a component population is known as *Standing Crop* (SC). Standing crop can be expressed in terms of no./unit area or in terms of "Biomass" (organisms' mass). Biomass can be measured as living weight, dry weight, CO_2 weight, calories, etc. for comparative purposes.

(b) Material removed: This includes the organisms removed from the ecosystem by emigration and material withdrawn as organic deposits (with reference to the yield to mass). For example rodents may be removed from a field for better agricultural yield. Hence this aspect has to be considered while calculating the secondary productivity of an ecosystem.

(c) Production rate: It is the rate at which growth processes occur progressively within the area. In areas where the production rate is more than the rate of destruction, a net increase in one or more of the components occurs as in eutrophication of lakes. Eutrophication refers to an excessive input of nutrients into aquatic ecosystem, which triggers on series of chain reactions. The excess of nutrient causes an accelerated growth rate of organisms, particularly in the upper layers of water. This results in a marked increase in biomass. The organisms, after their death, settle to the bottom as organic matter. As the rate of deposition of organic matter becomes more than the rate of its

decomposition, anaerobic conditions set in the bottom waters, turning the ecosystem into a stinking water body. An ideal situation would be always striking a balance between the rate of production and the rate of destruction.

Food Chain

The plants and animals of an ecosystem are linked together by their nutritional requirements. For example, the producers form the food for the herbivores, which in turn forms the food for carnivores. Individuals related in this manner constitute a food chain. The transfer of energy and nutrients from one feeding group of organisms to another in a series is called *food chain.* It is the sequence of eaters being eaten, or who eats whom.

The primary source of energy is the Sun, but the solar energy cannot be utilized by all groups of organisms. Green plants alone are able to trap solar energy, which they use to reduce carbon from carbon dioxide and this carbon forms carbohydrates, fats and proteins—the fuels of life. The energy trapped in these compounds is stored in the plants and forms the primary source of energy supply to all other living organisms. The autotrophic plants are, therefore, known as *producers.* In the animal community the plant eating animals or "herbivores" are known as the *primary consumers* which may include carnivores (flesh eating animals) or omnivores (feeding on mixed diet).

Each successive level of nourishment as represented by the links of the food chain is known as a *trophic level.* The producers of an ecosystem form the first trophic level, the herbivores form the second trophic level and the carnivores, the third trophic level. Energy is transferred from one trophic level to another.

The number of steps in a food chain is limited to four or five and at each step or transfer in the chain, a large portion of the potential energy is lost as heat. Hence the food chain with fewer number of links or transfers will prove to be the most efficient. Figure 3.2 shows a simple food chain in a pond.

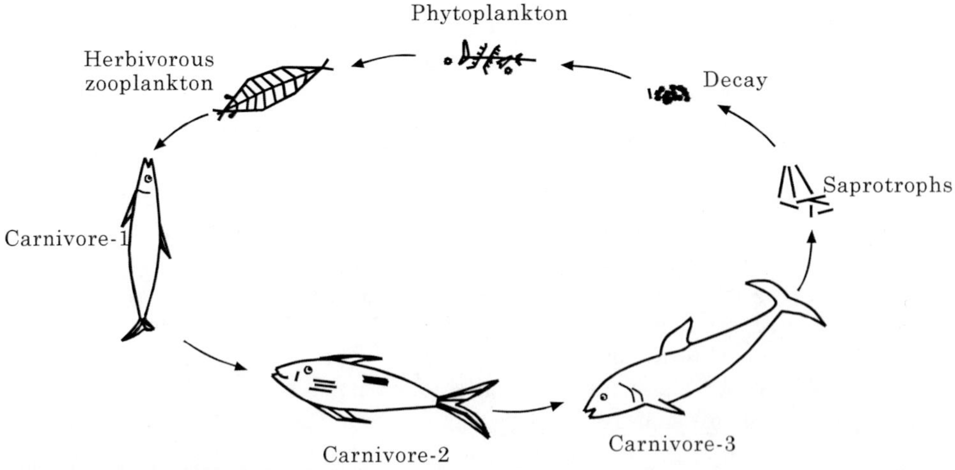

FIGURE 3.2 Simple food chain in a pond.

Types of Food Chains

In any ecosystem there are two major food chains, the *grazing food chain* and *the detritus food chain.* In most terrestrial and shallow water ecosystems, with their high standing crop and relatively low harvest of primary production, the detrital food chain is dominant. In deep-water aquatic ecosystems, with their low biomass, rapid turnover of organisms and high rate of harvests, the grazing food chain may be dominant.

(i) Grazing food chain: This food chain starts from living plants, goes through herbivores and ends in carnivores.

Plants → Herbivores → Primary Carnivores → Secondary Carnivores

Most of the ecosystems in nature follow the grazing type of food chains. This type of food chains depend directly on the autotrophs which capture the energy from solar radiation. A few chains are given below:

Grass → Grass hopper → Lizard → Hawk (Grassland)
Phytoplankton → Zooplankton → Fish → Snake (Pond)

(ii) Detritus food chain: It starts from dead organic matter and ends in inorganic compounds. There are certain groups of organisms which feed exclusively on the dead bodies of animals and plants. These organisms are called *Detritivores.* The Detritivores include algae, fungi, bacteria, protozoans, insects, millipedes, centipedes, crustaceans, worms and duck. These organisms ingest and digest the dead organic materials. Some amount of energy is trapped and the remainder is excreted in the form of simple organic compounds. These are again used by another set of detritivores until the organic compounds are converted into carobon dioxide and water. Detritus food chains do not directly depend on the autotrophs and solar influx.

Dead organic materials → Detritivores → CO_2 + H_2O

A good example of detritus food chain is found in mangrove ecosystem in which the leaves of plants like *Rhizophora* fall into the water which are eaten and re-eaten by a key group of small animals. These animals include crabs, insect larvae, shrimps, nematodes, etc. All these animals are detritus consumers. They ingest large amounts of plant detritus. These animals are in turn eaten by small game fish i.e. small carnivores which in turn serve as the main food for larger fish and fish-eating birds which form the top carnivores.

The two food chains cannot operate independently. They are interconnected at various levels. The stability of an ecosystem is directly proportional to the number of links. The detritus feeders obtain energy from the dead bodies of animals and plants which are components of the grazing food chains or vice versa. For example, in a pond ecosystem, earthworms belonging to detritus food chain are eaten by fishes, belonging to the grazing food chains.

Food Web

The food chains in nature never operate as an isolated, simple, straight-line sequences. Instead, they are interconnected with each other to form a net work called *food web.* (Figure 3.3) This is because each organism may obtain food

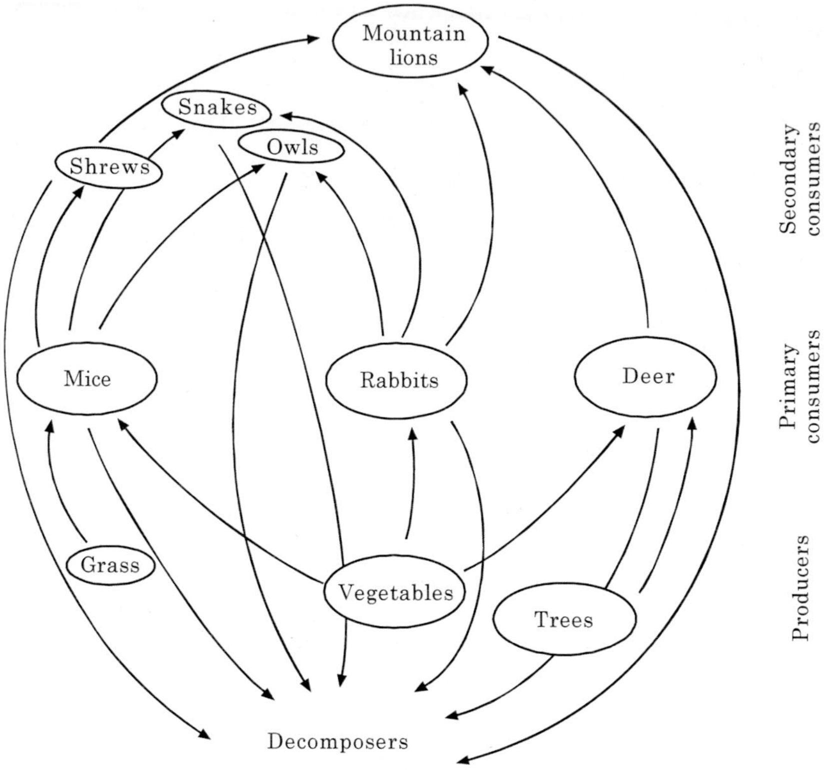

FIGURE 3.3 Food web.

from more than one trophic level. Otherwise each organism forms food for more than one organism of the higher trophic level. A sparrow may feed on primary producers (seeds of plants) or herbivores (leaf eating insects). Similarly in a grassland ecosystem, grass may be eaten by grasshopper, rabbit and mouse. Hence, in a food web, specific trophic levels cannot be assigned to species. Instead of this, the concept of trophic species was introduced. Species in a food web with the same diets and same predators form trophic species and are classified as basal, intermediate and top species. Basal species, which include both primary producers and detritus, are at the bottom of the food web and feed on no other species. The top species, occupying the apex of the food web, are one on which no other species feed. Intermediate species are neither basal nor top; they may feed on more than one trophic level. Predators are species that feed on other species of the web, and prey are species that are fed on by some other species.

Example: In a grassland ecosystem, grass is eaten by grasshopper, rabbit and mouse. Grasshopper is eaten by lizard which is eaten by hawk. Rabbit is eaten by hawk. Mouse is eaten by snake which is eaten by hawk. In addition, hawk also directly eats grasshopper and mouse, Thus, there are five linear food chains which are interconnected to form a food web. This is a very simple food web but in any ecosystem the food web is more complex with several interlocking chains. For example, in a grassland itself, in addition to hawk, there are many other carnivores such as vulture, crow, wolf, fox, man, etc.

Food webs can be separated into two large groups: those found in fluctuating environments characterized by variations in abiotic factors such as temperature, pH and salinity which are found in one type of environment. Food webs in constant environment are characterized by richness of greater species and more trophic interlinks. On the other hand, the food webs found in fluctuating environment contain only fewer trophic links. Thus the environmental conditions impose limitation and rigidity on the shape of the food webs.

Food chains tend to be short, rarely exceeding four links irrespective of the productivity of the ecosystem. The maximum number of such links in a food web is controlled by the second law of thermodynamics. The energy transfer from one trophic level to another is always inefficient. Thus the limitation of energy transformation forces a pattern to a food web. The further development of a food web is regulated by the properties of the existing food webs and invading species of an ecosystem. In any food web, there are four possible connections: (i) top and intermediate, (ii) different intermediates, (iii) intermediate and basal, and (iv) top and basal. The proportion in each is about the same for every habitat.

What happens if a species is removed or reduced in a food web? Theoretically there is a little effect if the species removed is a generalized prey species or a generalized predator species. If the species removed is a top predator, then the effect moves from the top species down through the intermediate to the basal. For example, if a top carnivore as tiger population is removed from a forest, the number of zebras, deers and other such plant eaters, which form food to the tiger, increases. These plant eaters will devour more vegetation leading to the depletion of foliage and plants which are producers. Thus the primary source of energy of the ecosystem is reduced affecting the whole energy budget and threatening its sustainability.

Significance of Food Webs

Food webs are very important in maintaining the stability of an ecosystem. The complexity of any food web depends upon the diversity of species and their interconnectivity. Diversity of species is based upon their food habits which would determine the length of the chain. More interconnectivity patterns in food webs suggest alternatives at different points of consumers in the chain. Such complex food webs are more stable than simple one for the reason that if a species is removed, its position is taken by another species of similar trophic status so that the ecosystem is maintained. For example, in a grassland ecosystem in the absence of herbivores, the plants will die due to overcrowding and competition for nutrients. When one type of herbivore is removed, the other types of herbivores increase in number and controls the vegetation. Thus each species of any ecosystem is kept under some sort of natural check so that the system may remain balanced.

Energy Flow

The transfer of energy from one trophic level to another trophic level is called *energy flow*. In an ecosystem the flow of energy is unidirectional. The study of energy transfer between different trophic levels in an ecosystem is known as *bioenergetics*.

The energy in an ecosystem is controlled by two laws of thermodynamics.

(i) Energy can neither be created nor destroyed.
(ii) Every transfer of energy is accompanied by its dispersion.

The source of energy for an ecosystem is the sun. Out of the total solar energy, 8 per cent strikes plants and only 80 per cent to 85 per cent of it is absorbed. Out of the energy absorbed, only 50 per cent is utilized in photosynthesis.

The efficiency of trapping solar radiation differs from one ecosystem to another and also from species to species. For example out of the total solar radiation that enters the atmosphere, aquatic ecosystem can absorb 0.2 per cent and terrestrial ecosystem 1 per cent. The efficiency of absorption varies from species to species.

The energy that is captured by plants is stored in the form of potential energy in food stuffs. These are producers and represent the first trophic level in the ecosystem. The energy stored by the producers is passed through food chain to primary, secondary and tertiary consumers. At each transfer a large proportion of potential energy is dissipated as heat.

Thus, the flow of energy in any ecosystem depends on the following factors:

(i) Efficiency of producers to trap solar energy and convert it into chemical energy
(ii) Use of chemical energy present in the producers by the consumers
(iii) Amount of energy assimilated during metabolic activities
(iv) Loss of energy in the form of unused energy, dead organisms and heat during respiration

The flow of energy through an ecosystem can be represented diagrammatically in a simplified manner.

Figure 3.4(a) indicates two aspects of energy flow in an ecosystem.

(i) The flow of energy is unidirectional.
(ii) There is a successive reduction in the energy flow at successive trophic levels.

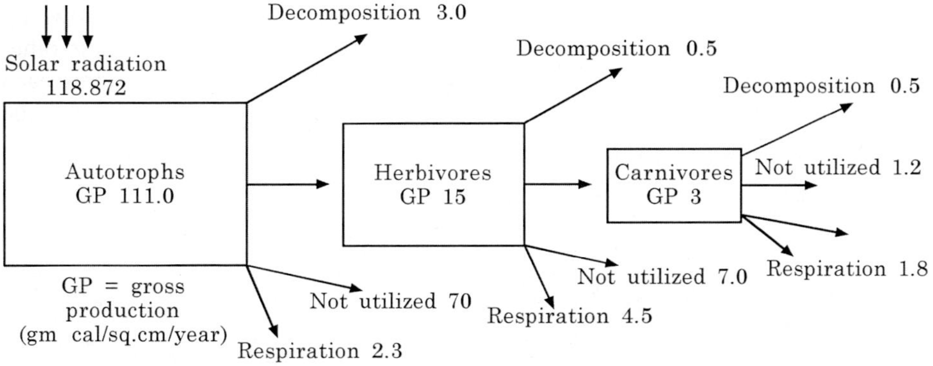

FIGURE 3.4(a) Energy flow model (lake) unidirectional.

At every trophic level certain amount of energy is not utilized or lost due to respiration and decomposition. Thus, longer the food chain, greater the loss of energy. The energy that is lost as heat will never return to the ecosystem. However, in such a one-way flow, the entire ecosystem would collapse if the primary source is cut off.

Y-shaped Energy Flow Models

In a Y-shaped flow model [Figure 3.4(b)] one arm represents grazing food chain and the other, detritus food chain. In the grazing food chain herbivores eat the plants and directly affect the plant population. Whatever they do not eat is left as organic matter (detritus) and forms food to the detritus feeders after decomposition. Similarly the undigested plant material (through faeces) is also available to the detritus feeders after decomposition. The Y-shaped model indicates that the two types of food chains are not isolated from one another and, in fact, one is a continuation of the other. For example, the dead bodies of small animals which were once a part of the grazing food chain become incorporated into the detritus food chain. The model is more practical than the single chain because of the following factors:

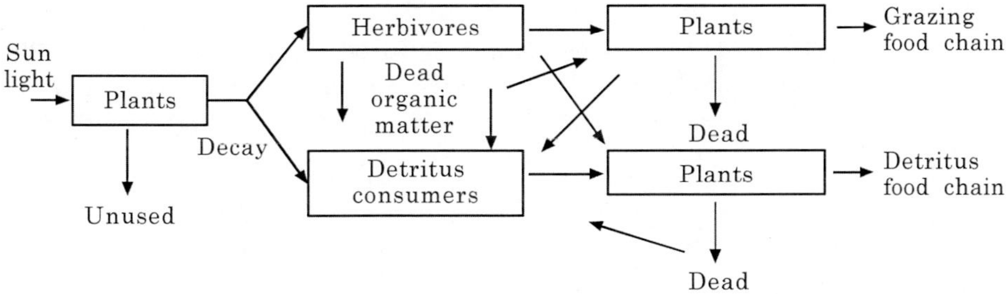

FIGURE 3.4(b) Y-shaped energy flow model.

(i) It confirms the basic stratified structure of ecosystems.
(ii) It separates the grazing and detritus food chains in time and space.

(iii) It shows that the microconsumers (absorptive bacteria) and macroconsumers (phagotropic animals) differ greatly in size related metabolism.

The above two models depict the basic pattern of energy flow in an ecosystem. In practice, under natural conditions, several trophic levels are inter linked and a complex food web (with multi channel energy flow) is formed. Thus, under field conditions the energetics of ecosystems can be measured only by estimating the primary and secondary production.

Ecological Pyramids

Ecological pyramids are the graphic representations of the number, biomass and energy of the successive trophic levels of an ecosystem. The number, biomass and energy of organisms gradually decreases from the production level to the consumer level. There are three types of ecological pyramids: pyramid of number, pyramid of biomass and pyramid of energy.

(i) The pyramid of number: The pyramid of number (Figure 3.5(a)) is the graphical representation of the number of individuals in various trophic levels of food chain per unit area at any given time. The number of individuals at the trophic level decreases from the producer level to the consumer level i.e. in an ecosystem, the number of herbivores is less than that of producers. Similarly, the number of carnivores is lesser than that of herbivores.

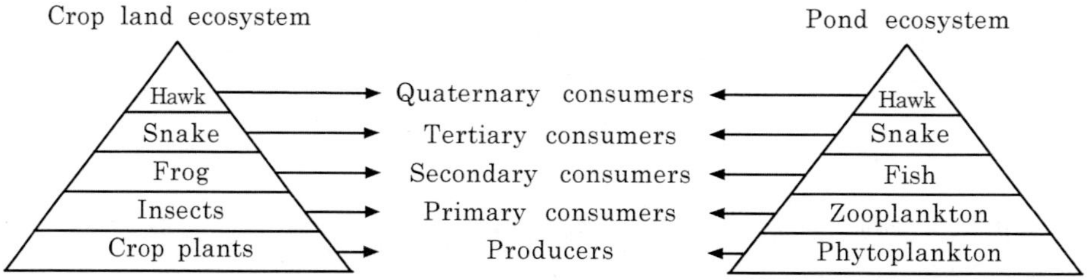

FIGURE 3.5(a) Pyramids of number.

In a crop land ecosystem, crop plants are more in number. The grasshoppers feeding on these plants are lesser in number. The frogs feeding on grasshopper are still lesser in number. The snakes feeding on frogs are still few in number.

(ii) The pyramid of biomass: Biomass refers to the total weight of living matter per unit area at a given time. In an ecosystem, the biomass decreases from the producer level to the consumer level.

Grass → mouse → snake → hawk
Plants → rabbit → fox → lion

(iii) The pyramid of energy: In an ecosystem, the energy flows from the producer level to the consumer level. At each trophic level

substantial portion of energy is lost. Hence the amount of energy decreases from the producer level to the consumer level. Since the flow of energy is always unidirectional, the pyramid of energy is upright. (Figure 3.5(b))

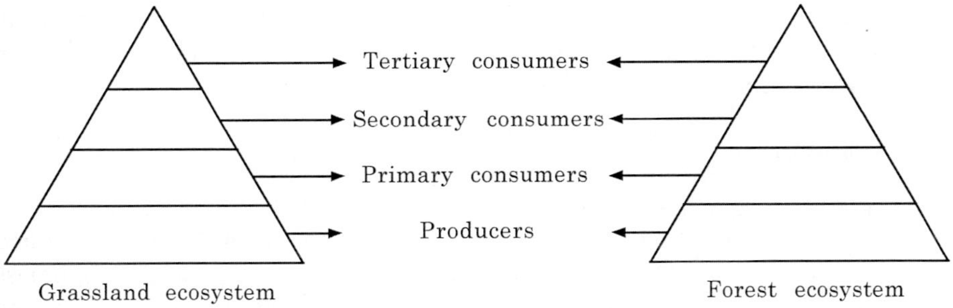

FIGURE 3.5(b) Pyramids of energy.

When the ecosystem contains lesser number and biomass of producers than those of consumers, the apex of the pyramid is directed downwards. Pyramids of this type are called *inverted pyramids.* Inverted pyramid of number occurs in a tree ecosystem. A single tree (producer) harbours many fruit eating birds (primary consumers) and these birds in their turn, host numerous parasites (secondary consumers). Inverted pyramid of biomass occurs in a pond or lake ecosystem. Here, the biomass of diatoms and phytoplankton is negligible as compared to that of either crustaceans or small fish.

Limitations of Ecological Pyramids

(i) The pyramid of energy has no appropriate or easy place for locating the decomposers and it does not allow to represent storage easily.
(ii) The pyramid of number equates all organisms as identical units.
(iii) The pyramid of biomass equates the unit weights of widely different groups.

Biogeochemical Cycles

Living systems (plants, animals and microbes) use and return the material to their reservoir sources in the biosphere (soil, water and air). Producers, consumers and decomposers play their roles to complete the cycle of events.

Water Cycle

Rainfalls on land and water bodies. It percolates into soil, flows on the land surface and joins oceans. Plants absorb water and release through evapotranspiration, solar heat also causes evaporation from water surfaces. Formation of clouds and again precipitation completes the hydrological cycle. Melting of ice contributes to water flow.

Carbon Cycle

Carbon element is essential for all biological systems. Carbon is present as carbohydrates (organic matter), CO_2, in carbonate rocks and fossil fuels. Combustion of carbon compounds and respiration of living units release CO_2 into the atmosphere. The process of photosynthesis prepares starch compounds (food products). Decaying vegetation and animal matter is again converted into carbonaceous deposits in course of long time interval.

Nitrogen Cycle

Elemental nitrogen is present in the atmosphere. Nitrogen is fixed for the plant utility by Rhizobium bacteria. Nitrosomonas help the chemical reactions for the formation of nitrites and Nitrobacter further converts it into nitrates. Denitrification is the process of reducing nitrates into nitrites, ammonia and nitrogen back into the biosphere. A number of microorganisms, fungi, algae help in completing nitrogen (nutrients, proteins) cycle.

Phosphorous Cycle

It is present as calcium phosphate in rocks. Plants take it in the form of orthophosphates and retain in their cells. Decay results in mineralization.

Sulphur Cycle

Oxidation and reduction of sulphur is responsible for the formation of SO_2, SO_3, SO_4, H_2S and S in the biosphere.

3.4 ECOLOGICAL SUCCESSION

In nature, environment is always kept changing over a period of time due to variations in climatic factors and activities of the species of communities themselves. These influences bring about marked changes in the dominance of existing community which is sooner or later replaced by another community at the same place. This process continues and successive communities develop one after another over the same area until a final community becomes stable for a longer period of time. This phenomenon is called *ecological succession.*

Ecological succession includes a series of changes that a biotic community undergoes in its maturation towards a stable or climax condition. All these changes are orderly, progressive and predictable.

Kinds of Succession

A. Ecological succession is of two types based on the nature of the habitat in which it begins.

(*i*) *Primary succession:* If succession proceeds from an area which has not been changed physically by organisms, it is called *primary succession.* **Examples:** (1) The colonization on a newly exposed island. (2) The development of a community on a rock. (3) The development of a community in a newly formed water body such as a lake or a pond.

(*ii*) *Secondary succession:* If succession starts on an area, previously colonized, and the soil is organically enriched, it is known as *secondary succession.* **Example:** The development of a community in cut-over forests, abandoned crop lands and the land that is deprived of communities due to volcanic eruptions, glacier movements and natural fire.

B. Ecological succession is of two types based on the type of organisms.

(*i*) *Autotrophic succession:* It begins in a predominantly inorganic environment, characterized by early and continued dominance of autotrophic organisms like green plants.

(*ii*) *Heterotrophic succession:* It begins in a predominantly organic environment, characterized by early dominance of heterotrophs i.e. fungi, bacteria and animals.

Trends in Succession

Ecological succession is a continuous process where one community replaces the other and this process includes the following four aspects:

(i) Change in species composition
(ii) Change in variety or diversity (diversity of a species tends to increase with succession)
(iii) Progressive increase in biomass (both living and dead organic matter)
(iv) Shift in community metabolism, such as a decrease in net community production (P) and a corresponding increase in community respiration (R)

In a young pond, $P/R > 1$, and in a stable pond, $P/R < 1$

Process of Succession

The whole process of primary autotrophic succession is completed through a number of sequential steps.

(*i*) *Nudation:* The process of formation of a bare area is known as *nudation* or *denudation.* There are several agents that cause nudation viz. physiographic, climatic and biotic agents.

(*a*) *Physiographic agents:* The action of wind, water, gravity, volcanic eruption, tectonic changes and glaciers result in erosion and deposition of soil. Erosion takes away fertile soil, organic matter, plants and animals with it. Deposition of soil destroys existing communities and leaves the surface bare for the succession to start.

(*b*) *Climatic agents:* Hail storm, drought and lightening destroy vegetation and animals even change the habitat.

(*c*) *Biotic agents:* Poisonous fumes from factories, deforestation by man and mining operations result in bare areas. In addition, overgrazing and epidemic spread of fatal diseases also make an area bare.

(*ii*) *Migration:* Movement of organisms into a bare area is called *migration.* Seeds and spores from adjacent areas are drifted into bare area with the help of wind drift or water flow. Mobility of animals due to their size, weight and adaptive characters leads to their migration into the bare area.

(*iii*) *Colonization:* Occupation of bare area by first or pioneer community is called *colonization.*

(*iv*) *Ecesis:* Establishment of a pioneer community is called *ecesis.* Such pioneers react with the medium (soil or water) and establish themselves. For this, the pioneer species should possess the following characteristics:

(a) It should be small in size and have low nutrient requirement.
(b) It should have slow growth rate.
(c) It should be gregarious.
(d) It should have interaction with others to withstand competition.
(e) It should be dynamic to face inter and intra specific competition.

(*v*) *Aggregation:* Aggregation is the final stage of invasion by pioneer group. The species which have successfully settled on the new area, reproduce and aggregate into large populations in the new area.

(*vi*) *Competition and reaction:* As the pioneer community becomes established, various species compete among themselves for space, light and nutrients. This competition results in a self-induced (autogenic) reaction on themselves and on the habitat. Communities which cannot withstand competition and changing environmental factors are replaced in succession by other communities till a climax community is established.

(*vii*) *Climax or stabilization:* This is the final stage in the process of ecological succession in which a community persists and is called a *climax community.* The climax community becomes more or less stabilized for a longer period of time and it can maintain itself in equilibrium with the climate of that area. If undisturbed, the climax community can maintain itself through years.

Succession is a property of the community driven by changes in attributes between youthful and mature systems. It eventually leads to the formation of an emergent entity with unique characteristics involving nutrient flow, biomass accumulation and species diversity. The pace of succession and the time taken to reach a climax stage varies with different climatic and physiographic situations.

3.5 TYPES OF ECOSYSTEMS

In nature, ecosystems operate as self-sufficient interacting units. They share a fundamental plan of their gross structure and function. However, they differ in their species composition and patterns of production, assimilation and transfer of energy depending on the habitat. Based on the habitat in which the biotic communities live, the ecosystems are classified into two major categories: Aquatic and Terrestrial.

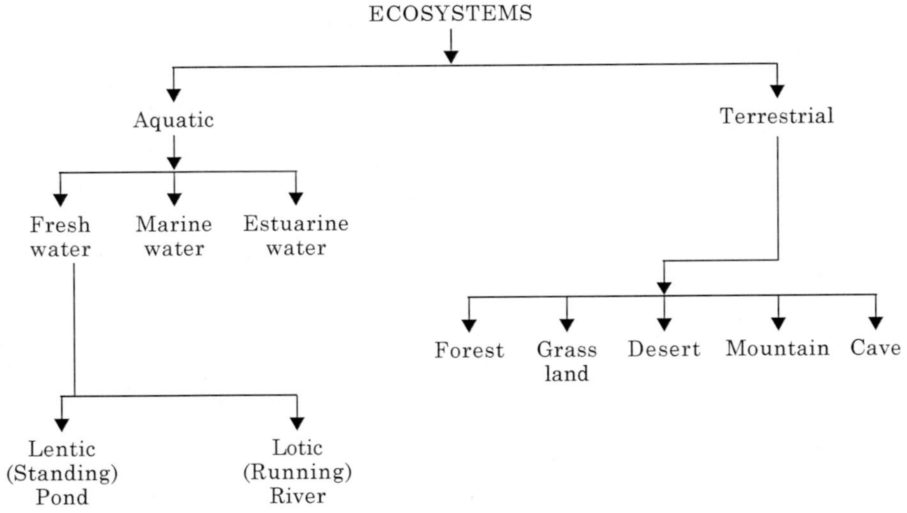

Aquatic Ecosystems

In all aquatic ecosystems, the medium in which the organisms live is water. It may be fresh water, salt water (marine) or a mixture of the both (estuarine).

1. Fresh Water Ecosystems

There are two categories of fresh water ecosystems: (a) Lentic and (b) Lotic. Fresh water ecosystems have low percentage of dissolved salts. They have fluctuating physical and chemical factors affecting the flora and fauna.

Lentic ecosystems

In the ponds and lakes the following three zones are evident depending upon the depth of water and the depth to which sunlight penetrates effectively.

(i) *Littoral zone:* It is the shallow water region with the light penetrating up to the bottom. Photosynthesis occurs actively. In natural ponds and lakes this region is occupied by rooted plants whose leaves float at the surface.

(ii) *Limnetic zone:* It is the open water zone which includes the vast expanse of water stretching upto the depth of effective light penetration. In this zone, the community is composed of only plankton and nekton. In small shallow ponds, this zone is absent. The total illuminated region of littoral and limnetic zones is called *euphotic zone.*

(iii) *Profoundal zone:* It is the zone of deep waters and the bottom. This zone is beyond the depth of effective sunlight penetration. It is often absent in ponds. Vegetation is absent and photosynthesis does not occur. Only saprotrophs are found here.

Pond Ecosystem

A pond is a good example of small ecosystem (Figure 3.6) that exhibits a recognizable unity both in function and in structure. To make a clear and detailed study of pond ecosystem, one should try to have information about producers (numbers, kinds, biomass and pigment densities), consumers (numbers, kinds and weights) and decomposers (kinds). This data provides an idea about the structure of the autotrophic and heterotrophic population.

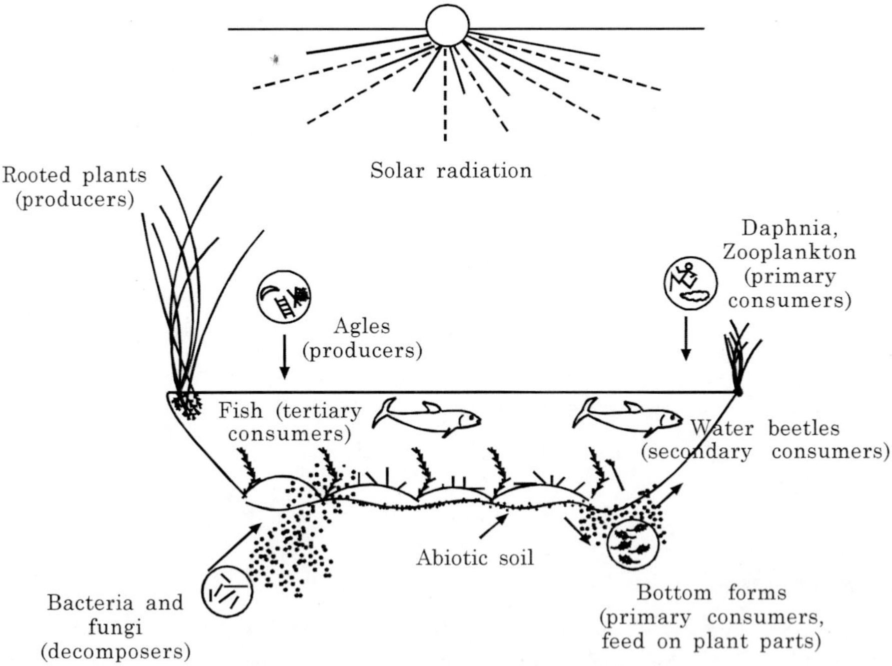

FIGURE 3.6 Pond ecosystem.

1. Abiotic Substances

Abiotic substances are basic inorganic and organic compounds such as water, carbon dioxide, oxygen, nitrogen, phosphorous, calcium, amino acid and humic acid.

(a) Inorganic substances: These include vital elements like C, H, O, N and P found both inside and outside the organism. They are involved in the exchange of materials between living and non-living parts and appear to remain in constant state of flux.

(b) Organic substances: Carbohydrates, proteins, lipids and humic substances are present. In addition to ATP which occurs only inside the living cells, humic substances are found outside the cells which form the major product of decomposition. DNA, chlorophyll, etc. are the other biotic complexes.

(c) Climate regime: It comprises of mainly temperature and other physical factors. Light and energy form the main non-living component for the growth of photosynthetic organisms (green plants and some algae).

However the rate of release of nutrients from the solids, the solar input and the cycle of temperature, day length and other climatic regimes are the most important processes which regulate the rate of function of the entire ecosystem on a day-to-day basis.

2. Biotic Components

Living components of the ecosystem constitute the biotic components and these include the following.

(a) Producer Organisms

(i) Phytoplankton: These are minute floating plants usually algae distributed throughout the pond as deep as light penetrates. They usually float with the current of the water. When in abundance, photoplankton give a greenish colour to pond water. These are very important in the production of basic food for the ecosystem. The phytoplankton of the pond usually comprises *Volvox, Ulothrix, Clostridium, Anabaena, Euglena* and *Ceratium.*

(ii) Filamentous algae: These algae will be floating in water and include *Spirogyra, Oedogonium, Nitella* and *Chara.*

(iii) Marginal and emergent plants: These include *Ipomea, Jussica* which are found floating on the surface and *Phragmites, Typha* and *Acorus* which are rooted plants or sedges.

(iv) Submerged plants: These are *Vallisneria* and *Potamogeton* which are rooted to the bottom. *Utricularia* and *Ceratophyllum* are rootless submerged plants.

(v) *Floating plants:* These include *Pistia, Lemna, Wolffia* and *Eichhornia.*

(b) Consumer Organisms

The macroconsumers represent animal fauna. These can be categorized into Primary consumers or herbivores, secondary consumers or carnivores and the tertiary consumers. The primary macro-consumers feed directly upon living plants or plant remains and are of the following types:

(i) *Zooplankton:* These microscopic animals *Dinoflagellates, Copepods, Heliozoans* and *Isopods* drift on the water surface currents.

(ii) *Nekton:* These are free swimming aquatic animals (insects and insect larvae).

(iii) *Benthos:* These are bottom dwelling forms (molluscs and annelids).

The secondary consumers or carnivores are predaceious insects and tertiary consumers are game fish.

(c) Decomposers

The aquatic bacteria, (flagellates and fungi) are distributed throughout the pond, abundant in the mud water. When temperature conditions are favourable, decomposition occurs very rapidly in the body of water.

Lake Ecosystem

A lake is defined as a large body of standing water which does not have connection with sea. It is deeper than pond. Lakes differ from ponds in retaining permanent water in the centre and by having sandy shore. It has stable environmental conditions.

Thermal stratification: The lake water exhibits a temperature gradient from surface to bottom. Different strata of water with different temperatures can be noticed. This is called *thermal stratification.* During summer three strata can be noticed. They are upper epilimnion, middle thermocline and lower hypolimnion. Epilimnion is the warm surface water. Here the temperature fluctuates with the atmospheric temperature. Water circulates in this layer. It has plenty of vegetation. The temperature will be 21°C to 25°C. The bottom layer is hypolimnion. It is the cold water layer. The temperature will be 5°C to 7°C. The water is stagnant and devoid of plants. The intermediate layer of water is called *thermocline.*

Based on the penetration of light, the water column of a deep lake is divided into three zones. The euphotic zone is a lighted zone and it extends to a depth of 80 metres. In the disphotic zone light is highly modified and it extends from 80 to 200 metres. Aphotic zone is without light and it extends beyond 200 metres.

Classification of lakes: Lakes are classified into three types, on the basis of productivity.

(a) *Oligotrophic lakes:* Oligotrophic lakes are characterized by the following features:

(i) They have great depths
(ii) They are poor in organic materials
(iii) Oxygen is abundant
(iv) They have low nutrients
(v) They are poor in fauna and flora

(b) *Eutrophic lakes:* Eutrophic lakes are characterized by the following features:

(i) They are shallow
(ii) They are rich in phosphorus
(iii) They are rich in organic materials
(iv) They have high fertility
(v) They are rich in flora and fauna

(c) *Dystrophic lakes:* Dystrophic lakes are characterized by the following features:

(i) They may be shallow or deep
(ii) They are rich in P, N_2, and Ca
(iii) They have high organic content
(iv) Oxygen content is very low or absent
(v) They are poor in fauna and flora

The biotic component of lake is similar to that of a pond.

River Ecosystem

Rivers are lotic (running water) systems. Lotic habitat is characterized by motion of water. When water particles move parallel to one another, the flow is called *laminar flow.* In rivers with turbulent flow, the erosive power and the forces at the water-sediment interface are very high.

(i) Land-water interchange is greater in lotic habitat.
(ii) Oxygen is abundant in running water system.

Zonation of river: The deeper part of the river is called the *flowing water zone.* The shallow water areas with an irregular substrate of rocks are named the *riffle zone.* In the zone of deeper water, the velocity of current is low.

Biotic components of river: Producers of river include diatoms, blue green algae, green algae, water moss, etc. Consumers of river include nymphs of dragon flies, may flies and stone flies, beetles, water skaters, snails, clarias, frogs, etc.

Stream Ecosystem

Organisms (periphyton) capable of adhering to an exposed surface (rocks, limbs, etc.) are found in the upper reaches of a stream. The periphyton includes filamentous green and blue-green algae and various sesile invertebrates, including larvae of insects. Further downstream, floating and emergent vegetation maybe found along with sesile invertebrates and those that burrow in the softer substrate,.

Chemically the upper reaches of lotic environment are rich in oxygen, but as the water moves downstream and becomes more sluggish, the oxygen level tends to drop. Because of the continual addition of nutrients and detritus enroute, nutrient levels tend to be higher downstream. In small streams, the major source of nutrients is from external ecosystem.

Biotic Components: The plants and animals inhabiting streams have unique adaptations in order to adjust with the current and flow of water.

(a) *Producers:* Phytoplankton is absent in rapidly flowing waters but mosses and filamentous green algae grow attached to the rocks. But in the pool zone of slow flow of water, phytoplankton is present which is transported from adjacent standing waters.

(b) *Consumers:* In rapidly flowing water zooplankton is absent. Fishes can swim against the water current. Some animals live in burrows to withstand the force of water flow. In pools where water flow is slow, zooplankton is abundant in addition to insects, snails, cat fish, crocodiles and turtles.

2. Salt Water (Marine) Ecosystems

Marine habitat is the largest one in the biosphere and it covers an area of 3,62,000,000 sq. km. approximately equal to 71 per cent of the earth's surface. It supports a vast assemblage of varied forms of life. The reason for the greater representation of organisms in the sea is due to the remarkable stability and uniform conditions of the sea waters.

Characteristics of a marine ecosystem are as follows:

(i) *Volume:* It is the largest habitat.

(ii) *Continuity:* Marine habitat is characterized by continuity. All the seas of the world are connected with one another.

(iii) *Depth:* Marine habitat has great depths. The average depth is 4000 metres.

(iv) *Salinity:* Salinity is very high. It is about 35 per cent. The salinity is due to the abundance of chlorides of Na, K, Ca and Mg.NaCl forms about 27 per cent. As many as 49 elements are found in sea water. The salinity remains stable in most of the oceans.

(v) *pH:* Sea-water has more hydroxyl ions (OH) and hence it is alkaline in its nature. The pH is 8 to 8.3.

(vi) *Currents:* Sea-water is in continuous circulation due to the currents of water caused by wind, temperature and gravitational force.

(vii) *Tides:* Tides are regular rise and fall in the level of the sea, caused by the attraction of the sun and the moon.

(viii) *Pressure:* Pressure increases as the depth increases.

(ix) *Temperature:* The temperature of an ocean ranges from about 2°C in the polar seas to 32°C in the tropics.

(x) *Light:* Light is a significant factor in regulating the pattern of the distribution of marine plants and animals. According to the depth of penetration of light three layers are recognized in the marine habitat. *Euphotic Zone* is the upper layer of water. It extends from surface to 80 metres. It is the lighted zone. Producers are abundant in this layer. *Disphotic Zone* is the middle layer. This zone extends from 80 metres to 200 metres. This zone is characterized by the presence of diffused or modified light. Producers are rare in this layer. *Aphotic zone* extends below 200 metres. Light and producers are completely absent in this layer.

Stratification of Marine Habitat: The marine habitat exhibits clear horizontal stratification as pelagic region and benthic region. Pelagic region includes the entire water body, further sub-divided into neritic and oceanic zones. Benthic region consists of the floor of the sea. It is subdivided into littoral zone and deep sea region.

The Ocean Ecosystem

Oceans, constitute one of the great reservoirs of living things and of the essential nutrients needed by both the land and the marine organisms. As in other ecosystems, the life in the ocean also depends upon light. The energy of light is stored in the form of carbon compounds, which are used for the substances and energy of all organisms. The life in the ocean is affected by important physical factors like currents, tides, depth, temperature, light penetration and salinity.

Oceans are divided into several general regions depending on the depth of the bottom (Figure 3.7).

(i) *Neritic zone:* Most continents are surrounded by a continental shelf, a more or less flat plain under 200 metres of water but in some places much less. The shelf and the water over the shelf constitute the neritic zone. The neritic zone can be further sub-divided into supratidal (above the high tide mark), intertidal (between the high and low tide lines) and subtidal regions.

(ii) *Oceanic zone:* Beyond the neritic zone is the oceanic zone. This is the part of the ocean which forms the "blue water". The layers upto 3000 m depth are the abyssal region. No light ever penetrates this region. The temperature is virtually constant at about 3°C.

(iii) *Euphotic zone:* The upper (100 m) part of the ocean, into which enough light penetrates for photosynthesis, is known as the euphotic zone. The floor of the ocean is thrown into gigantic ridges and valleys.

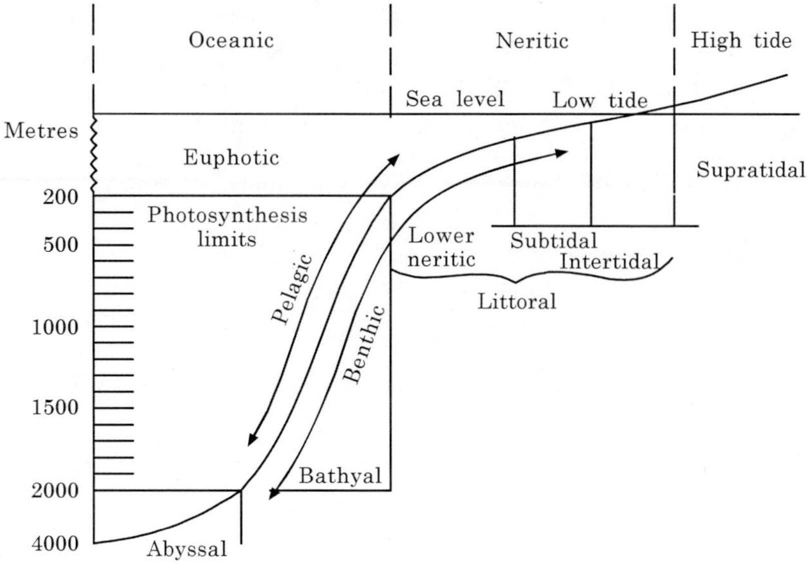

FIGURE 3.7 Zones in sea.

The following are the important biotic components of an ocean ecosystem (Figure 3.8).

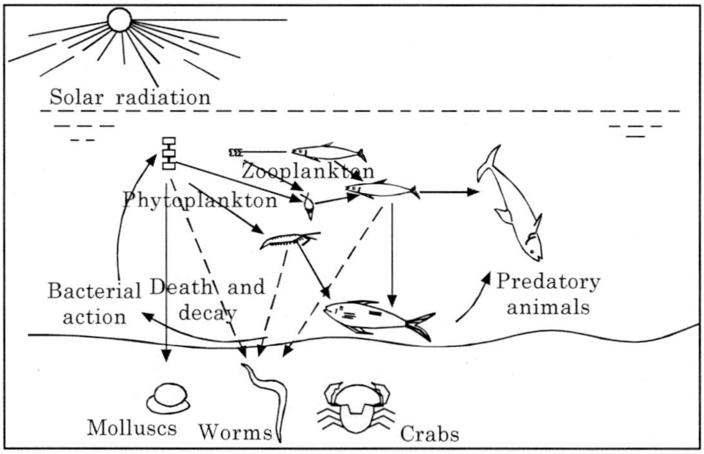

FIGURE 3.8 Ocean ecosystem.

(i) *Primary producers:* Phytoplankton, mostly diatoms, green flagellates are the primary producers in the open ocean. In addition to this, a few microscopic algae, floating weeds and green, brown and red algae form the basis of life in sea. All of them are present in photic zone where sufficient light is available.

(ii) *Consumers:* The primary consumers in an ocean ecosystem are crustaceans, molluscs and herbivorous fishes. Fishes like cod, haddock, halibut, etc. are the secondary carnivores. All of these fishes are bottom feeders. The fishes in the sea ecosystem have a tendency to remain together and to undergo seasonal migrations. This is due to their physiological and reproductive compulsions.

(iii) *Decomposers:* Bacteria are the decomposers, which bring about decomposition of living organisms and the dead bodies of the plants and animals. Then the decomposers convert the material to suitable nutritive forms that could be consumed by green plants once again.

3. Estuarine Water Ecosystem

Most of the rivers generally join the sea. The meeting place of the river and the sea is called an *estuary*. It is a narrow passage or the mouth of a river or lake where the tide meets the current.

Characteristics of an estuary

(i) It is the ecotone of marine and freshwater habitat.
(ii) The sediments carried by the river, accumulate at the river mouth.
(iii) Water level in the estuary fluctuates.
(iv) During high tide more sea water enters the estuary and during low tide the sea water recedes.
(v) Salinity is intermediate between fresh water and sea water.
(vi) Water currents are present in the estuary due to interaction of flowing streams, oceanic tides and wind.
(vii) Estuarine temperature fluctuates daily and seasonally.
(viii) The mixing of sea and fresh waters produces turbulance and eddies. The eddies contain most of the nutrients of the estuaries. Hence they are called nutrient traps of self enriching system.
(ix) Estuaries have a high silt content. It reduces the penetration of light.

Biota of estuaries

The organisms of estuaries are classified into Oligohaline, True estuarine, Euryhaline marine, Stenohaline marine organisms and Migrants.

(i) Oligohaline organisms are freshwater organisms. which cannot tolerate variations in salinity of more than 0.1 per cent. They are not found at the head of the estuary.
(ii) True estuarine organisms are restricted to estuaries. They have adaptations to a wide range of salinity.
(iii) Euryhaline marine organisms extend their distribution from the sea to the upper reaches of the estuary. They can tolerate salinity as low as 15 per cent. These organisms form the majority of total estuarine biota.
(iv) Stenohaline marine organisms live on open seashore and at mouths of estuaries. They do not enter estuaries below salinities of 25 per cent.
(v) Migrant organisms spend only a part of their lives in estuaries.

The biota of estuaries is varied and represented by animals and plants that can tolerate fluctuations in salinity. These organisms spend only a part of their lives in estuaries. Some examples of biota are *Euglena*, *Volvox*, *Cyclops*, hermit crab, fishes like *Tilapia* etc. *Macrophytes* (plants) are weeds, marsh grasses and sea grass.

Terrestrial Ecosystem

Terrestrial habitat refers to the land where the organisms live. Land is characterized by a variety of climates, diversity of abiotic factors and heterogenity of biotic communities.

Characteristics of terrestrial habitat are as follows:

(i) There is scarcity of water and dryness also prevails.
(ii) Temperature fluctuations are more.
(iii) The intensity of light is high.
(iv) Atmospheric air is the source of O_2.
(v) O_2 and CO_2 are provided at a constant rate.
(vi) Land is discontinuous, with a number of geographical barriers in the form of mountains, valleys, lakes, rivers, seas, etc.
(vii) Soil contains all the necessary nutrients for the production of energy.

Ecologically, the terrestrial habitat is subdivided into a number of sub units called *biomes*. A biome is defined as a major terrestrial community characterized by distinctive plants and animals.

Examples: 1. Forest biome, 2. Grassland biome, 3. Tundra biome, 4. Coniferous biome, 5. Cave biome, and 6. Desert biome.

Characteristics of Biomes

Each biome has a climax community which is dominant. The climax community forms the matrix of the biome. The biome is named after the climax community. For example, in a grassland biome, grasses form the climax community and in a forest biome trees form the climax community. In addition to the climax community, each biome has many intermediate communities. Thus a biome consists of a special combination of complex of communities. The communities of a biome are maintained under the climatic conditions of the region. There is no clear-cut demarcation between adjacent communities. The adjacent communities blend in the meeting place to form an ecotone. A biome is different from an ecosystem, in that the ecosystem consists of both biotic and abiotic factors. But a biome includes only plants and animals.

The Forest Ecosystem

Forests are the regions where there is sufficient quantity of moisture and a moderate temperature. Moisture, temperature, rainfall, snow, wind, velocity, etc. are more or less stable and govern the forest environment. Angiosperms and gymnosperms are the forest trees. The days are cooler and nights are warmer in forests.

Forests may be of three types depending upon the climate, soil conditions and available water content.

(i) *Coniferous forests (taiga):* They lie adjacent to tundra biome at high altitude and extends across both North America, Sweden, Finland and Russia. Siberia is rich in coniferous forests. It is characterized by needle like leaves of evergreen trees, (poplar, birch and pine trees) and animals like black bear, porcupine, snow shoe hare, lyrix, red fox, wolf, red squirrel, northern flying squirrel, owl, woodpecker, etc. The presence of green plants increases the annual rate of primary production of this biome. The temperature is cold during winter. Growing season is very short. Summers are of very short durations and rainfall is low. The soil is thin, acidic and poorly drained. Numerous lakes are associated with these forests.

(ii) *Deciduous forests:* They cover Europe, part of Japan, Australia and top of South America (tropical and temperate areas). The deciduous trees shed off their broad leaves, become bare in winters and develop again in spring. The rainfall in this biome is abundant (75 to 150 cm) and evenly distributed throughout the year. Temperature exhibits seasonal variations and is moderate round the year. The season varies from north to south.

The deciduous forests are rich in both invertebrate and vertebrate fauna. Invertebrates include earthworms, snails, millipedes, insects (caleoptera and orthoptera), etc. whereas vertebrates include amphibians (newts, toads, frogs, salamanders, etc.) reptiles (turtles, snakes, lizards), birds (great horned owl, hawks, woodpeckers, etc.) and mammals (opposum, pigs, squirrels, bear, deer, raccoon, mountain lion, red and grey foxes, etc).

(iii) *Tropical rainforests:* The communities of this biome are very much diversified with evergreen plants having broad leaves. The annual rainfall is about 150–200 cm and is the most important ecological factor. The temperature of forests is uniform throughout the year and the stratification is maximum. The trees form three layers namely scattered, tall emergent trees; second canopy layer; and understory streams. The climate is warm and humid.

Tropical forest areas are grouped into three regions namely, Amazon and Orinico basins in South America, Congo, Niger basins of central and West Africa and Indo-Boraeo-New Guinea regions. In India, tropical rain forests stretch along western ghats as well as western and eastern Himalayas.

The outstanding features of plants in tropical rainforests are the presence of tall, broad-leaved, evergreen trees, presence of epiphytes, abundance of lichens, large number of trees and uniformity of leaves. The trees are very tall (upto 60 metres) and below them are bamboos, shrubs, ferns, etc.

The forests are very rich in vertebrate: the tall trees and other vegetation provide home to these animals. Frogs, (*Rhacaphorus*) toads, chameleons, lizards, snakes, woodpeckers, spider monkeys, howler monkeys, squirrel monkeys, bats, sloth, tigers (*Panthera tigris*), elephant (*Elephas maximus*), deer, chital deer (*Axis axis*), swamp deer, and gaur (*Bibos gaunes*) are the common vertebrate fauna represented in tropical rain forests.

In India about 22.7 per cent (74.5 million) hectares of the total land area is covered with forests. They fall into the categories of the alpine forests, sub-alpine forests, the Himalayan forests, the moist tropical forests, the dry tropical forests and the thorn forests.

The Grassland Ecosystem

Natural grasslands occur where the rainfall is intermediate between the desert and forest land (25 to 75 cm) and occupies about 19 per cent of the area of land (20,800,00 sq. km.). The principal grassland ecosystem are great plains of Canada, United States of America, South Argentina to Brazil and South Asia to Central Asia. Nilgiri Hills constitute the grassland in India. Greater variations in temperature, moisture, wind, intensity of sun light have been observed in this ecosystem.

The grassland commonly builds an entirely different type of soil as compared to that of forest, though both start with the same minerals. Grasses are short-lived and a large amount of organic matter is added to the soil. The decay is rapid and results in the formation of humus. The dark grassland soils are well suited for growing food plants.

(i) *Abiotic substances:* All the elements which are requried for the composition of living organisms are stored in the earth and its atmosphere. The plants get carbon, hydrogen, oxygen, nitrogen, phosphorous and sulphur and supply them to the consumers as carbon dioxide, water, nitrates, phosphates and sulphates. In addition to these, calcium, sodium, potassium, iron, copper, zinc, cobalt, etc. are also supplied by the soil.

(ii) *Primary producers* are the scattered shrubs, trees and grasses.

(iii) *Herbivores* (mostly insects, reptiles and birds) are the consumers. Common insects found are termites, *Eoccinella, Leptoconia* etc. Reptiles and millipedes are also common consumers found in grasslands.

(iv) Bacteria and fungi are the commom *decomposers*. For example, *Aspergillus* and *Cladosporium* are found in abundance. Their activity is maximum in wet monsoon season.

The Desert Ecosystem

Desert biomes occur in the regions with less than 10 cm of annual rainfall or sometimes in hot regions, where there is more rainfall, but evenly distributed in the annual cycle. The deserts cover about 17 per cent of the total land.

Important deserts are the Somoran desert (America and Mexico), the Sahara (North Africa) and Thar desert (India) and Central portion of Australia.

Four types of plants are found in desert ecosystem.

(i) The annuals (cheat grass) which avoid drought by growing only when there is adequate moisture.
(ii) The desert shrubs with thick leaves which survive by becoming dormant before wilting occurs. In cooler deserts, shrubs develop deep roots to absorb moisture and in such cases, the leaves and stems remain green during summers.
(iii) The succulents which store water in their tissues.
(iv) Microflora such as mosses, lichens, blue green algae, etc. that remain dormant in soil but respond to cold or wet periods.

Some animals, reptiles and insects adapt themselves to the deserts. Their skin becomes thick and they excrete uric acid in the form of dry pellets. Mammals (with the exception of a few rodents and camels) are poorly adapted for desert life. Camel can store water in its body. The desert animals have various adaptations in their morphological and physiological aspects.

Water and high temperature are the main limiting factors in desert. Thus, the productivity of a desert is directly related to the amount of annual rainfall.

Mountains and Caves

Mountains

Earth is moving and changing by few mm per year through its crustal plates, so that mountain ranges are continually being built up or worn away. Valleys are formed due to the erosion of soft rocks down the slope.

Observations on the mountain ranges are as follows:

1. less oxygen in the atmosphere
2. snow on the top of high mountains
3. temperature falls by 1°C for every 15 m elevation
4. less vegetation
5. strong winds, more clouds and rainfall
6. hot days and cold nights

Rivers contribute to the formation of fertile soil and development of various crops and plantations. Forest products, minerals, iron ore, coal, etc. are also available in the mountain ranges with dense forest areas in the lower elevations.

Plants and animals have extra protections depending on the surrounding environmental conditions.

Many communities are very isolated as transport facilities are minimal.

Caves

Cracks, crevices, potholes or a network of passages are formed either by acidic water dissolving carbonates (limestone) or due to decaying vegetation in different rocks. Water seeps through these cracks in rock formation giving way to underground streams, tunnels and caves. Material deposition causes ridges. When the roof portion collapses, caves extend. Sometimes caves are interconnected. The most favourable situation is in between a permeable limestone and impermeable rock deposits. Animals take shelter in the mountain caves.

SUMMARY

Ecosystem is a location (habitat) for the living organisms for maintaining a balanced relationship with biotic and abiotic components of environment for their food and energy requirements. Food material is prepared (by producers for consumption) by a simple photosynthetic activity from simple substances like water, and carbon dioxide in favourable conditions (temperature, light, oxygen, nutrients). Again the material is decomposed back into the environment for recycling. Nutrients and energy moves from one group of feeding organisms to another known as *food chain.* Each organism may obtain its food from various sources and these food webs help in maintaining the stability of an ecosystem. Energy also moves (unidirectional) from one trophic level to another. Environment always changes with time due to various factors like climate, species interaction and the development process of communities is known as succession always leading to a mature systems. Ecosystems are broadly classified into aquatic (river, lake, sea) and terrestrial (forest, grassland, desert) units. Each one exhibits recognizable units both in function and structure and operates on a self sustaining basis.

Commercial activities and environmental pollution are the real threats to the ecosystem sustainability.

KEYWORDS

Abiotic components, aquatic and terrestrial ecosystems, consumers, decomposers, ecology, ecological pyramids, ecosystem, energy flow, food chain, material cycling, producers, succession.

SOLVED QUESTIONS

1. **Comment on species interactions.**

   (i) Different species cannot coexist for a long period if they depend on the same source. Species use the environment in different ways. This segregation reduces competition and permits coexistence.

(ii) When resources deplete, species compete and the fittest survives.
(iii) Organisms adjust to prevailing environmental conditions or migrate to suitable places.
(iv) Interactions among various organisms like symbiosis, antagonism, cannibalism, parasitism, predation, amensalism, affect the rate of growth of populations.

2. Give two indices for the evaluation of environmental stress.

(i) Weight of grass (biomass) on land
(ii) Excess of eutrophication in lakes

3. What are the threats to ecosystem sustainability?

Commercial activities that man engages need raw materials from agricultural, mineral, marine, forest and animal products. The following are of some major environmental concern:

(i) Rapid use of renewable resources beyond their natural capacity to regenerate
(ii) Reduction of biodiversity and forest cover
(iii) Escalating use of energy resources
(iv) Unplanned infrastructural growth causing stress on fragile ecosystems
(v) Use of chemicals and their disposal
(vi) Global warming and other serious concerns
(vii) Population growth and poverty

4. How are biotic and abiotic elements related?

Solar radiation is the only input. All other elements have cyclic relationships. Hydrological cycle illustrates the abiotic–biotic relationships. The carbon cycle explains the abiotic–biotic relationship.

The biotic components are linked in many ways through trophic levels to form food chains and food webs. They maintain relationships with members of their own species, with members of other species and community relationships. Living and non-living things are inseparably interrelated and interact with each other.

5. Explain the terms 'dynamic' and 'stable' with reference to ecosystems.

Distribution of energy flow and material changes from place to place and time to time. It is a functioning live system. Any change sets up corresponding cyclic changes in other components sometimes till the change pervades the whole environment. While change due to natural forces is sequential, regulated and orderly, man made changes are uneven leading to maladjustment between man and nature.

All changes in nature are in a state of balance and adjustment. Biosphere has a special self purification process of replenishment. The system is said to be stable if the depletion and degradation of resources is minimum and if it is sustainable.

REVIEW QUESTIONS

1. Define ecosystem and describe its components with suitable examples.
2. Describe the structure and functions of a lentic ecosystem.
3. In nature, grazing and detritus food chains are operated simultaneously. Explain.
4. Describe the process of ecological succession.
5. One of the major functions of an ecosystem is the flow of energy through its components. Explain.
6. Write an essay on the characteristics of marine ecosystem.
7. Describe the structure and functions of an estuarine ecosystem.
8. Explain the significance of forests as major biomes with reference to productivity.

OBJECTIVE TYPE QUESTIONS

1. Which of the following is a biotic component of an ecosystem?

   (a) Fungi (b) Ecesis (c) Temperature (d) pH

2. In an ecosystem, the flow of energy is

   (a) Bidirectional (b) Cyclic (c) Unidirectional (d) Multidirectional

3. The science of control is called

   (a) Genetics (b) Cybernetics (c) Ecology (d) None

4. Deciduous forests are present in

   (a) Tropical and temperate zones
   (b) Tropical and sub-tropical zones
   (c) Temperate and sub-tropical zones
   (d) Temperate and polar zones

Fill in the blanks with suitable words:

5. The pyramid of ______________ is always upright.
6. Inverted pyramid is generally found in ____________.
7. In complex ecosystems a _____________ degree of species diversity is found.
8. In an ecosystem the flow of energy is __________ and the flow of nutrients is _______________.
9. Distinguish between the following:

   (a) Autotrophs and heterotrophs
   (b) Food chain and food web
   (c) Biome and ecosystem

CHAPTER 4

BIODIVERSITY AND ITS CONSERVATION

Contributed by Dr. (Smt.) A. PRIYAMVADA DEVI,
St. Ann's Degree College for Women, Hyderabad

TOPICS COVERED

- Introduction
- Composition of Biodiversity
- Factors Affecting Diversity
- Value of Biodiversity
- Biodiversity at Global Level
- Threats to Biodiversity
- Endemism and Biodiversity
- Biogeographical Zones of India
- Hot Spots of Biodiversity
- Conservation of Biodiversity

4.1 INTRODUCTION

More than 8 lakhs of varying organisms such as plants, animals, insects, birds, reptiles and microbes support each other and live together. They are not evenly distributed in the world as each classified group requires different environmental conditions for their growth, multiplication and welfare.

Biodiversity refers to the variety and variability among living organisms and the ecosystem complexes in which they occur. It includes diversity of forms right from the molecular unit to the individual organism and then on to the population, community and ecosystem. Thus, biodiversity is the degree of variety of nature. These biological variations in nature, over time and space, form the basis of evolutionary processes. New elements of life by—mutation, natural or artificial selection, speciation or artificial breeding, biotechnology or ecological manipulations can be viewed as additions to biodiversity. On the other hand extinction of species, disappearance of ecological associations and loss of genetic variants in the existing species can be considered as a loss of biodiversity.

4.2 COMPOSITION OF BIODIVERSITY

Biodiversity has three major components: genetic diversity, species (organismal) diversity and ecosystem (ecological) diversity. These three are intimately interlinked and actions at any one level will have impact on the other levels of the hierarchies.

The composition and levels of biodiversity can be represented as follows:

Genetic Diversity	Ecological Diversity	Species (organismal) Diversity
Populations	Biomes	Kingdoms
↑	↑	↑
Individuals	Landscapes	Phyla
↑	↑	↑
Chromosomes	Ecosystems	Families
↑	↑	↑
Genes	Habitats	Genera
↑	↑	↑
Nucleotides	Niche	Species
	↑	↑
	Populations	Sub-species
		↑
		Populations
		↑
		Individuals

Genetic Diversity (Diversity within species)

Animals and plants (eukaryotes) are composed of basic units of structural organization called *cells*. Each cell has a vital component, the nucleus, which in its turn contains a substance called *deoxyribonucleic acid* (DNA). DNA is the carrier of genetic material that is passed on from generation to generation. In order to carry out the life processes such as growth and reproduction, the cells of the organisms undergo repeated divisions. During such divisions the DNA of the nucleus organizes itself into rod like structures called chromosomes. Chromosomes are the physical basis of heredity. The number of chromosomes vary in different species of plants and animals but remains constant and characteristic for a particular species. For example: the number of chromosomes (2n) present in man is 46, fruit fly is 8, and onion is 32.

The grades of structural organization of chromosome are as follows:

1. Sugar + Phosphate + Nitrogen base = DNA
2. DNA + Histones (Proteins) = Histone fibre (Nucleosome)
3. Histone fibre $\xrightarrow[\text{on condensation}]{\text{on coiling}}$ Chromatin fibre
4. Chromatin fibre $\xrightarrow[\text{and coiling}]{\text{on condensation}}$ Chromatid
5. 2 Chromatids + Centromere → Metaphase chromosome

i.e. DNA → Histone fibre → Chromatin fibre → Chromatid → Chromosome.

Deoxyribo Nucleic Acid (DNA) is a double helical structure [Figure 4.1 (a) and (b)] formed of polynucleotide chains. Sugar and Phosphate form two

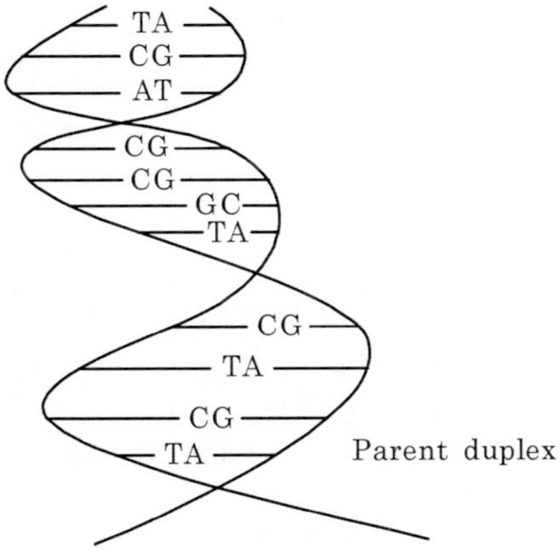

FIGURE 4.1(a) DNA structure.

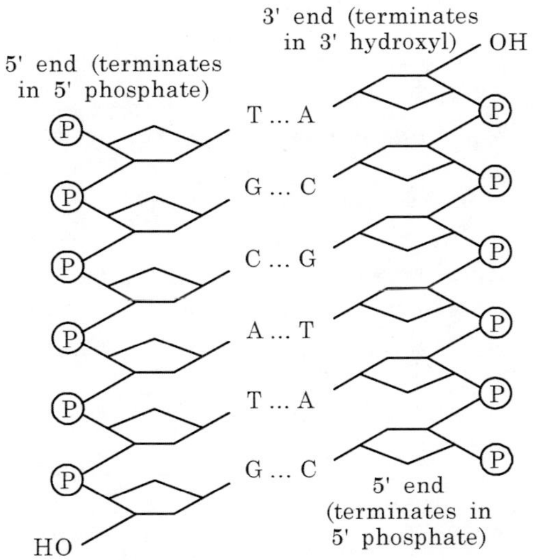

FIGURE 4.1(b) DNA structure.

parallel chains and nitrogen bases (Purine = Adenine and Guanine; Pyrimidines = Cystine and Thymine) are attached to them at right angles like the rungs of a ladder. DNA combines with histones (proteins) and undergoes coiling and condensation to form a chromatid. Each metaphase chromosome is formed of two chromatids that are attached to a centromere.

The function of DNA is to synthesize polypeptide chains (proteins). **A specific region of DNA coding for one polypeptide chain is called a gene**. Thus, gene is the fundamental unit of inheritance and all living organisms store their biological information (messages) in the form of genes. **Genetic diversity refers to variations in individuals of a species, attributable to the differences in their genes**. It is expressed as a change in the physical characters within a population, leading to polymorphism (occurrence of several structural forms). Genetic diversity is a response shown by the organisms of a species to meet the challenges of the environment. In the population of human beings, occurrence of different races such as negroes, whites and mulattoes and others with all intermediate shades of skin colour (the Indian population) is an example of maintenance of genetic diversity. Similarly, occurrence of red, pink and white flowers of evening primrose (*Oenothera lamarkiana*) is an expression of genetic diversity.

Genetic diversity of a population arises by the combined and independent effect of mutations and genetic recombinations.

(i) ***Mutations:*** Mutations are sudden, stable and heritable changes in the genetic material of the individuals. They occur spontaneously in nature with no assignable cause. But mutations can be induced by agents (mutants) like UV rays, X-rays, certain chemicals as nitrous oxide, acridine dyes that damage DNA.

Mutations are of the following two types:

(a) Gene mutation or a change in the gene
(b) Chromosomal aberrations or change in the structure and number of chromosomes

(ii) ***Genetic recombination:*** When cells undergo division (meiosis), a reciprocal exchange of genes occurs between chromosomes by a process called *crossing over*, resulting in genetic recombination. In addition to this, recombination of genes also occurs independently due to random segregation of chromosomes during cell division. The genetic recombinations alter the physical traits of organisms causing variations in a population.

Significance of Genetic Diversity

(i) Nature operates selection on organisms and in this process the fittest of all, will have more chances of survival. If a species has more genetic diversity, it provides ample opportunity for natural selection to act on it and maintain polymorphism.

(ii) The amount of genetic variation is the basis for 'speciation' (evolution of new species). It has a keyrole in the maintenance of diversity at both species and community levels. The total genetic diversity of a community will be greater if there are many species compared to a situation where there are only a few species.

(iii) Genetic diversity within a species, often increases with environmental variability.

Species Diversity

Species diversity refers to the variety of species and abundance of individual organisms of each species within a region. Earth has about 12.25 millions of species of which only two millions have been classified. The major goal of modern conservation biology is to ensure the maintenance of high species diversity in biosphere.

There are several indices of species diversity. The simplest of all is to count the number of species per unit area. This is called *species richness*. The species richness increases from the poles to the equator. This could be attributed to a number of biotic and abiotic factors on a local scale and to the high rate of evolution on a large scale. Species richness affects several aspects of community function. In general, the communities with species richness perform better than those without richness. Species-rich communities are more resistant to invasion by exotic species (non-native organisms). Species diversity of a habitat depends on its species richness. However, the 'evenness' or 'equitability' of richness depends on the abundance of the individual in each species i.e. a quantitative measure. However, the quality of species richness is determined by the degree of taxonomic diversity of the community. "Taxonomic diversity refers to the occurrence of species belonging to distantly related groups of plants and animals in a community".

Ecosystem Diversity

The number of ecosystems can vary within a geographical area. Grasslands, evergreen forests, deserts, lakes, wetlands and oceans are the major ecosystems where species live and evolve. The number of ecosystems present in a region is also a measure of its biodiversity.

Ecosystem diversity has thrce perspectives:

(a) Alpha diversity: It is the diversity of organisms exhibited within a community. A combination of species richness, evenness and taxonomic diversity is used to represent alpha diversity.

(b) Beta diversity: To keep pace with the changes in the habitat, the composition of the species of a community also changes. Thus communities are dynamic and ever changing. The rate of replacement of species along a gradient of habitats brings in diversity between

communities. This is called *beta diversity:* There are differences in the composition of species in communities along the environmental variations of moisture, altitude etc. The beta diversity of a region is directly related to the heterogeneity of the habitats and communities.

(c) *Gamma diversity:* It is the diversity of the habitats over the total landscape or geographical area.

The diversity of an ecosystem describes the number of niches, trophic levels and various ecological processes that sustain energy flow, food webs and the recycling of nutrients. Studies on temperate grasslands have shown that diverse communities are functionally more productive and stable even under environmental stresses such as prolonged dry conditions.

4.3 FACTORS AFFECTING DIVERSITY

Several factors influence the species diversity of different habitats. Some of them are as follows:

(i) **Time:** New communities arise when new habitats are formed and such communities will have only few species. More number of species would be added subsequently in the course of time by various processes.

(ii) **Extreme habitats:** Habitats that have extreme conditions of cold, heat, acidity or pollution tend to have relatively few species. This is due to the fact that many organisms lack genetic capabilities for evolving the ability to tolerate a particular kind of extreme habitat.

(iii) **Resource diversity:** The diversity of groups of organisms is directly related to the degree of diversity of resources. For example, a forest with more vegetation could support more species of birds than a grassland. The transition zone between two adjacent ecosystems is called an *ecotone*, and in these regions a high degree of species diversity is observed.

(iv) **Productivity:** In nature, more species can exist in areas of high productivity than in those of low productivity. In a productive area, some features of the ecosystem, such as a certain kind of food will be abundant that becomes a useful resource for one or more species.

(v) **Climatic stability:** Areas of stable and predictable climate will have more species than those of unpredictable climate.

(vi) **Predation:** Predators may keep the numbers of certain of their prey low enough that competitive exclusion does not occur. In the rocky intertidal zone of the Olympic peninsula in Washington, starfish is the top carnivore that preys on mussels and barnacles. When the starfish was experimentally removed, several species disappeared and a mussel that is a preferred food of the starfish came to dominate the area.

(vii) **Disturbance:** A disturbance is an interruption to the settled state. A fire, a flood or the outbreak of a disease can be considered as a disturbance to the ecosystem. At high levels of disturbance, many species are lost and only those that can tolerate the disturbed conditions survive. Similarly at low levels of disturbance, the most nearly climax species may come to dominate to a degree at which many of the other species are eliminated.

4.4 VALUE OF BIODIVERSITY

Each and every component of the nature has its unique, immense value (unknown wealth) for more than one reason, viz. ecological, practical, aesthetic, ethical and social. Everything is related (i.e. interconnected) to everything else in the ecosystem. Even small things are very important in maintaining the stability of an ecosystem. System collapses without biodiversity. Resources should be conserved without fail. The value of biodiversity can be listed as follows:

(i) Natural areas are places for scientific research and education in all disciplines as Geology, History, Literature and Life Sciences.

(ii) Many natural areas are irreplaceable, especially in a short period of time. To produce a forest, starting with a barren ground, would take several hundreds of years through ecological succession.

(iii) When the species with their unique genotype (genetic constitution) are preserved, they form a source of disease-resistant new crops, improved breeding varieties, biological pest control and sewage processing. For example, in Asia four varieties of disease-resistant rice are produced from a single wild species (*Oryza nivara*).

(iv) Biodiversity is the source of food for man. Man occupies the apex of the trophic levels and gets his food from all other organisms. But he would hardly acknowledge this and rarely appreciate the contribution of his fellow species. As many as 80,000 species of plants are edible for human beings. Similarly herbivores and carnivores also form the food source to man.

(v) Several important drugs and medicines are prepared from plant-based substances (Table 4.1). About 25 per cent of the drugs in pharmacy are derived from only 120 species of plants. But all over the world, traditional medicines make use of thousands of plant species. Animals and microbes also form a good source of pharmaceutical products. The (UNDP) development programme estimates that the value of pharmaceutical products derived from plants, animals and microbes of developing world is about $30 billion per year.

(vi) Biodiversity has great aesthetic value. Landscapes are the delightful sites of nature in their pure form. Ecotourism, bird-watching and wildlife observation camps are the recreational activities associated with life in nature.

Table 4.1 Some natural medicinal products

Product	*Source*	*Use*
Penicillin	Fungus	Antibiotic
Bacitracin	Bacterium	Antibiotic
Tetracycline	Bacterium	Antibiotic
Erythromycin	Bacterium	Antibiotic
Digitalis	Foxglove	Heart stimulant
Quinine	Cinchona bark	Antimalarial
Diosgenin	Mexican yam	Birth control drug
Cortisone	Mexican yam	Anti-inflammatory drug
Morphine	Poppy	Analgesic
Vinblastine and Vinaristine	Periwinkle Plant	Anti-cancer drug
Reserpine	Rauwolfia	Hypertension drug
Cytarabine	Sponge	Leukaemia cure
Bee venom	Bee	Arthritis relief
Allantoin	Blowfly larva	Wound healer

(vii) Plants and animals are recognized in the society for national pride, cultural heritage and religious belief. In several places of our country some plants like Tulsi (*Ocimum sanctum*) and Pipal (*Ficus religiosa*) and animals like Cobra (Naga raja), monkey (Hanuman) and eagle (Garuda) are considered sacred and worshipped.

(viii) Biodiversity needs to be preserved for a valid cause, as genetic material is the end product of millions of years of evolution.

(ix) Ecosystems and their constituent organisms offer several benefits to mankind which include maintenance of the gaseous composition of the atmosphere, biogeochemical cycling, soil formation, watershed management, pest control and pollination.

(x) Human beings are a part of nature rather than apart from it. They are members of biosphere, not the owners. In its most ethical sense underlying biodiversity is that the concern for fellow species makes a man superior to beasts.

4.5 BIODIVERSITY AT GLOBAL LEVEL

Biological diversity is the result of interaction between climate, organisms, topography, parent soil materials, time and heredity. It is estimated that 5 to 30 million species of living forms exist on our earth of which only 1.5 million have been identified. They include 3,00,000 species of green plants and fungi, 8,00,000 species of insects, 40,000 species of vertebrates (fishes, amphibians, reptiles, birds and mammals) and 3,60,000 species of microorganisms.

The tropical forests are regarded as the richest in biodiversity. More than 50 per cent of the species on the earth live in moist tropical forests which are only 7 per cent of the total land surface. This maintenance of high degree of biodiversity by tropics is due to favourable conditions for evolution and low rates of extinction.

Biodiversity at National Level (India as a Megadiversity Nation)

The Indian region (80–30°N and 60°–97.5°E), having a geographical area of 32,87,263 sq. km. is quite rich in biodiversity with a good percentage of endemic flora and fauna. This richness in biodiversity is caused due to the combination of a variety of climatic conditions and ecological habitats. These vary from the humid tropical western ghats to the hot desert of Rajasthan, from the cold desert of Ladakh and icy mountains of the Himalayas to the warm coasts of peninsular India.

India has over 1,15,000 species of plants and animals which are already identified and described. In addition to this, the country is an important gene bank and source for the origin of over 167 cultivated plant species and domesticated animals. These species arose in the country and subsequently got spread throughout the world. To name a few among them: rice, sugarcane, jute, mango, citrus, banana, many species of millets, several cucurbits, some ornamental orchids, several medicinal and aromatic plants. This country is also a secondary centre of diversity for grain of amaranthus, maize, red pepper, soya bean, potatoes and rubber plant.

There are several examples of plant germplasm from India making significant contributions worldwide for plant improvement. To cite two examples:

(i) From the wild variety of rice of Kerala, nearly twenty different disease-resistant varieties are produced which are cultivated in different parts of the world.

(ii) From the wild musk melon of India, powdery mildew-resistant genes are contributed to the cantaloup (musk melon) industry (14,000 horha) of California of US.

Plant Wealth of India

India has about 45,000 species of plants which form 15 per cent of the world flora. There are 15,000 species of flowering plants of which 35 per cent are endemic. Among monocotyledons, out of 558 genera occurring in the country, 22 are exclusively endemic. The north-eastern region could be considered to be the 'treasure house of orchids' in the country, bearing about 675 species out of 1,000 species available in Indian peninsula and against 17,000 species of the world. The important Indian orchids are *Paphiopedilum fairieyanum, Cymbidium aloiflium* and *Aerides crispum.*

Many of the plant species have lot of economic importance. About 1,000 wild edible plant species are widely exploited by native tribes. These include 145 species of roots and tubers, 521 of leafy vegetables/greens, 101 of buds and flowers, 647 of fruits and 118 of seeds and nuts.

Animal Wealth of India

India is rich in its fauna bearing nearly 75,000 species of animals. Insects form the major group amounting to 80 per cent of the fauna of the country. The country has, to its credit, a high rate of endemism, 33 per cent of reptiles and 62 per cent of amphibians are endemic to India against the world percentage of 6.9 and 4.32 respectively. Further, there is a wide diversity in domestic animals such as buffaloes, goats, sheep, pigs, poultry, horses, camels and yaks. The western ghats in peninsular India which extend into the southern states are a treasure house of species diversity. It is estimated that almost one third of the animal varieties found in India are present in the western ghats of Kerala alone.

India has a coastline of 7,000 km with a shelf zone of 4,52,460 sq. km. and extended economic zone of 20,13,416 sq. km. Sea-weeds, fishes, crustaceans, molluscs, corals, reptiles and mammals are abundant in the coastal water and deep water of the seas. Owing to the richness of plant and animal species (Table 4.2), India is recognized as one of the megadiverse country of the world.

Table 4.2 Number of plant and animal species in different groups recorded in India

Plants and bacterial group	*No. of species*
Angiosperms	17,500
Gymnosperms	64
Pteridophytes	1,100
Bryophytes	2,850
Lichens	2,000
Fungi	14,500
Algae	6,500
Bacteria	850
Animals	
Mammals	390
Aves	1,232
Reptiles	456
Amphibians	209
Pisces	2,546
Protochordates	119
Molluscans	5,070
Arthropods	68,389
Protozoans	2,577
Other invertebrates	8,329

4.6 THREATS TO BIODIVERSITY

The major threat to biodiversity is extinction of species. Extinction or elimination of species is a normal process in nature. As a part of evolutionary change, species die out and are replaced by others, often their own descendants. In undisturbed ecosystems, the rate of extinction appears to be about one species lost per decade. In this century, however, the disturbances caused by human beings have accelerated the rate leading to the elimination of several kinds of plants, animals and microbes, threatening the very existence of life on our planet.

Himalayan dragon fly, coconut crabs and snails of Andamans, Indian python, Himalayan quail, North India's white-winged duck, Nilgiri's langur, lion-tailed macaque, malabar's large spotted civet are disappearing slowly.

Natural Causes of Extinction

Studies of the fossil record suggest that more than 99 per cent of all species that ever existed are now extinct. Most of those species were gone long before man came into existence. Mass extinctions have wiped out vast numbers of species periodically (Table 4.3). At the end of Cretaceous period dinosaurs disappeared along with at least 50 per cent, the then existing species. A greater disaster occurred at the end of Permian Period about 250 million years ago when 90 per cent of species and half of all families died out over a period of about 10,000 years which makes only a meagre fraction on the geological time scale. Several theories suggest that these catastrophes were caused by climatic changes, perhaps triggered when large asteroids struck the earth. Many ecologists of today are of the opinion that the global changes caused by the enhanced release of "green house" gases into the atmosphere could have similar dangerous effects.

Table 4.3 Mass extinctions

Geological period in geological time	*Time in million years (before present)*	*Effects (extinction)*
Quaternary	Present	1/3 to 2/3 of all species become extinct if the present trend continues.
Cretaceous	65	17 per cent of families, 50 per cent of species including dinosaurs but not mammals.
Triassic	210	23 per cent of families, 50 per cent of species.
Permian	250	54 per cent of families, 90 per cent of species.
Devonian	370	19 per ccnt of all families.
Ordovician	444	25 per cent of all families.

Anthropogenic Threats to Biodiversity

Between 1600 and 1850, human activities appear to have eliminated two or three species per decade, which is about double the natural extinction rate. But, in the last 150 years, the extinction rate has increased to thousands per decade. If the present trend continues, by the end of twenty first century 1/3 to 2/3 of all the existing species would become extinct.

There are a number of anthropogenic factors leading to the extinction of species and consequent loss of biodiversity. Some important factors among them are: habitat loss, deforestation, fragmentation, introduction of exotic species, over-exploitation of natural resources, poaching, disturbance and pollution, control of pest and predators, agricultural practices and other factors.

(i) Habitat Loss

Disturbance to natural habitat created by man is the largest single cause of loss of biological diversity. Over the past 10,000 years, billions of hectares of forests, woodlands and grasslands have been converted to commercial forests, crop lands or grazing lands.

Temperate broad-leaved forests are the most completely human dominated biomes. The climate and soils that support such forests are especially congenial for human occupation. Temperate grasslands have been widely converted to agricultural lands. Similarly tropical forest losses are accelerating in many parts of the world. Wet lands and estuaries occupy relatively little area world wide, but they support high diversity and biological productivity. Wetland loss is a key factor in biodiversity loss.

In India, out of 3 million sq. km. habitat of original wild life, only 20 per cent is remaining. Human impact on habitats takes two main forms: *conversion from one type of habitat to another* and *modification of conditions within a habitat.* The area of crop land and irrigated crop land has increased and the area of forests and grasslands has decreased by about 15 per cent in the last century.

(ii) Deforestation

Forests are the most species-rich terrestrial habitats. Loss of forests is rapid in the developing countries. Deforestation affects biodiversity through destruction of habitat, isolation of fragments of formerly continuous habitat, 'edge effects' between forested and deforested areas and a variety of other mechanisms. The consequences of tropical deforestation for biodiversity can be categorized into three types:

(i) Reduced diversity of species and genes:
   - (a) species extinctions
   - (b) reduced capacity to breed improved crop varieties
   - (c) inability to make some plants as economic crops
   - (d) threat to the production of minor forest products

(ii) Changes affecting local and regional ecosystems:
   - (a) soil degradation
   - (b) changes in water flow from catchment areas
   - (c) changes in buffering of water flows by wet land forests
   - (d) increased sedimentation of rivers, reservoirs etc.
   - (e) possible changes in rainfall characteristics

(iii) Changes affecting the global ecosystems:
   (a) reduction in carbon stored in the terrestrial biota
   (b) increase in carbon dioxide content of the atmosphere
   (c) changes in global temperature and rainfall patterns by green house effect
   (d) other changes in global climate due to changes in land surface processes

(iii) Fragmentation

Fragmentation is the reduction of habitat into smaller and more scattered patches. It reduces biodiversity because many species such as bears, tigers and lions require large territories for their living. Other species such as, the birds that live in the interior of forests can reproduce successfully only in deep forests but not in the edges or close to human settlement. Fragmentation also divides population into isolated groups and such groups are vulnerable to catastrophic events such as a single storm or an outbreak of a disease. Very small populations may not have enough breeding adults to be viable even under normal circumstances. Species diversity is a balance between colonization and extinction rates. When the population was having only less than 10 breeding pairs of birds, 39 per cent of them became extinct over a period of 80 years. At the other extreme, when the breeding pairs were between 100 to 1000, only one species became extinct.

With the loss and fragmentation of habitat, the species must either adapt to the changes, move elsewhere or may succumb to predation, starvation or disease and eventually die. In our country, several rare species of butterflies are facing extinction with the rapid destruction of the habitats in Western Ghats.

(iv) Introduction of Exotic Species

Aliens, introduced into habitats where they are not native, are called *exotic species.* When exotic species are introduced into a habitat they enter into competition with native species for food and shelter. This is one of the major threats of biodiversity, next to habitat loss.

Species are introduced into alien habitats by people for a number of reasons. Such introductions can lead to severe disruption of ecological communities.

Exotic species have wide range effects on ecosystems affecting both the structure and the function. They may eliminate native species directly through animal predation or the browsing effects of herbivores. To quote a few examples:

- Where ungulates were introduced into the temperate grasslands in Australia, South America and western North America, the composition of the native species has undergone massive changes.
- The introduction of Nile perch (fish), *Lates niloticus*, into the Lake Victoria eliminated about 200 to 300 endemic cichlid species of fish.

- The exotic plant fungal pathogen Cryphonectria parasitica eliminated American chestnut tree Castanea dentata from eastern USA.
- When Baltic salmon (fish) was introduced into the rivers of Norway for aquaculture purposes, the Atlantic salmon were eliminated as the former carried a pathogenic parasite *Gyrodactylus salaris* (flatworm) which turned to be deadly for the latter.
- Asian long horned beetle, *Anoplophora galabripennis* was introduced into US in 1999 through logs and wooden packing crates from China. This is a wood-eating insect, which burrows into the trunks of living trees as maple, oak and poplars. It makes tunnels and weakens the trunk, leading to the death of the trees. Thus, populations of these shade giving trees were greatly reduced in size.

Introduction of goats into Galapagos islands has resulted in the death of a large number of two important reptiles, the land iguana and the giant tortoise which can live for 100 years. Goats ate voraciously the plants which form the food of tortoise and iguana.

In India, the plant *Lantana camara* has invaded many forest lands and is strongly competing with the native species.

(v) Poaching of Wildlife

Plants and animals are the organisms whose degree of protection depends on the value they represent for human beings. This specific anthropocentric view leads directly to the subordination of biological diversity and to its sacrifice inspite of the understanding of the necessity for conservation.

Poaching is indiscrimate and illegal killing of animals for pleasure or profit. This, in combination with illicit trade of wildlife species, is threatening the diversity of species. Killing of fur-bearing animals such as leopard, tiger, jaguar, crocodile, snake and deer is on high degree for their skin, teeth, nails etc. The skin is used for making fur coats, shoes and bags.

The African ostrich has nearly reached extinction due to the demand for its feathers in making hats, hand fans and other luxury items. The worldwide demand of ivory took the toll of elephants too. The whales were endangered due to the demand of whale products such as lipsticks, face creams, perfumes, oils and pet foods.

The trade of wildlife presents a dilemma. It places a market value on wild species, encouraging sustainable exploitation rates which form a valuable source of income for local people. Convention on International Trade in Endangered Species of Wild Flora and Fauna (CITES), which prohibits all trade in a species, has listed about 600 threatened species of animals and plants that might be affected by the trades. If a species is 'rare', its value increases dramatically manifolds making it more vulnerable to extinction.

The illegal trade of rhino's horn which is used for medicinal and ornamental purposes is responsible for the drastic decline in the number of black rhinoceros over most of its range in Africa during the last 15 years. The exploitation of parts of tiger for the Asian medicinal market is the current biggest threat to the survival of tiger which is already under severe pressure from habitat loss. The gall bladders of Asian bear are priced 18 times the price of gold on weight basis. The bear is already facing the problem of a drastic reduction in the numbers and populations due to great habitat destruction and human encroachment. Similarly a single tiger's skeleton fetches $ 255 to $ 3400 and the bones $ 15 to 200 per kg to the poachers and middle men in India. Tusks of elephants and sandalwood are the most lucrative of the wildlife trade. The profits in wildlife smuggling are enormous.

Overharvesting of wildlife due to unawareness of ecological balances in nature and selfish commercial motives are the most dangerous threats to biodiversity. A visit to any tourist coastal area in India like Kanayakumari or other southern parts, reveals that everyday a number of marine invertebrates including corals, echinoderms, molluscs and crustaceans are collected, killed, treated with chemicals and sold to tourists. Most remarkable among them are corals and sea urchins.

Millions of birds, reptiles, amphibians and mammals are often illegally traded as pets to developed countries.

Among plants, wild Ginseng is eliminated in many areas because of the Asian demand for its roots, which are used as an aphrodisiac and folk medicine.

(vi) Disturbance and Pollution

Communities are affected by natural disturbances such as fire, tree fall and defoliation by insects. Man-made disturbances such as release of synthetic compounds, radiations into nature or spillover of oil into the sea causes pollution. These impacts lead to qualitative changes in the habitat ultimately reducing the biodiversity. Seals, dolphins and cranes died in large numbers due to toxic chemicals like DDT, PCBs, Pb, Acid rain, pesticide, pollution of soils and eutrophication of water bodies reduces the diversity of species.

(vii) Control of Pests and Predators

Predator and pest control measures generally kill predators that are a component of balanced ecosystem and may also eliminate several non-target species.

(viii) Collection for Zoos and Research

Animals and plants are collected throughout the world for zoos and biological laboratories for study and research in science and medicine. Rabbits, mice and monkeys are sacrificed for research and earthworms, frogs, garden lizards, prawns and fish are dissected in educational institutions. A strategy can be planned carefully for the use of animals only at a higher level of education, as scientific and medical research where sacrificing animals for a larger cause is inevitable.

(ix) Other Factors

In addition to the causes described above, some other factors may also contribute to the extinction of plant and animal species:

(a) **Distribution range:** The smaller the range of distribution, the greater the threat of extinction.

(b) **Degree of specialization:** The more specialized an organism is, the more vulnerable it is to extinction.

(c) **Position of the organism in the food chain:** The higher the position of the organism in the food chain the more susceptible it is to extinction.

(d) **Reproductive rate:** Large organisms tend to produce fewer offspring at widely spaced intervals. The populations of such organisms are more prone to extinction compared to those with high rate of reproduction in short intervals.

4.7 ENDEMISM AND BIODIVERSITY

Endemic species are those that are confined to a particular geographical area or ecological unit. Therefore endemism signifies unique biodiversity. Endemics are not randomly distributed across the earth but tend to be clustered. Majority of the centres of endemism have been identified on the basis of endemic flowering plant (flora) and large-bodied terrestrial vertebrate (fauna). Very high endemism of terrestrial taxa is recorded in the mid latitudes of the southern hemisphere, particularly with the Mediterranean type of climate. The lowest levels are recorded in northern hemisphere which was glaciated during the Pleistocene period. Centres of endemism for marine algae are concentrated in mid latitude areas with the highest percentage recorded along the temperate southern Australian coast. The percentage of invertebrate endemics near the shore have increased southward and reached a peak in the subantarctic zone.

A study of the endemic species forms the basic inventory work towards understanding biodiversity of that area. Endemics are highly vulnerable to anthropogenic disturbance and other forms of environmental change and are often indicators of biodiversity-rich locations.

Endemic Species of India

India harbours a number of endemic plant and animal species. Among the animals, reptiles and amphibians show a high degree of endemism. There are 18 species of endemic reptiles and 110 species of endemic amphibians. Eight of the amphibian species are exclusively Indian variety, not found anywhere in the world, viz. *Bufoides* (toad), *Ranixalus* (frog) and *Melanobatrachus*. Among reptiles, gharial is endemic and endangered, 55 species of birds are endemic to India. Majority of them are distributed along the mountains of eastern India, western ghats and Nicobar and Andaman islands. Their distribution is dependent on the average annual rainfall. Indian skimmer, Hodgson's frogmouth and some varieties of pheasants are endemic species of birds. There are only 44 species of mammals endemic to India. To name a few among them are lion tailed macaque, Nilgiri langur, Indian lion, Indian wild ass and brow-antlered deer. Majority of the endemic species are endangered and need to be protected from extinction.

Among the plants, 140 genera with more than 5,000 species are identified as endemic. 33 per cent of flowering plants and 18 per cent of all other plants found in India are endemic. North-east India, western ghats, north-western and eastern Himalayas, Andaman and Nicobar Islands are the rich sites of endemism. Exceptional diversity is found in Taxa such as ferns (Pteridophytes) with 900 recorded species and orchids with 1,082 species.

Endangered Species of India

The World Conservation Monitoring Centre has recorded that 533 animal species (mostly vertebrates) and 384 plant species (mostly flowering plants) have become extinct since the year 1600. The rate of extinction is maximum in islands. It is estimated that the earth may lose upto 50 per cent of the species by the end of the 21st century, if the current rate of loss continues.

Susceptibility to Extinction

The characteristics that make a species susceptible to extinction are as follows:

- Large body size (elephant, lion and Bengal tiger)
- Low reproductive rate and small population size (blue whale and giant panda)
- Feeding at a high trophic levels in the food chain (Bengal tiger and bald eagle)
- Fixed migratory routes and habitats (blue whale and whooping crane)
- Localized and narrow range of distribution (woodland caribou and many island species)

The IUCN Red List Categories

The International Union for Conservation of Nature and Natural Resources (IUCN) has a catalogue of taxa that are facing the risk of extinction. This is called *Red List* or *Red Data Book.* The objectives in preparing the Red List are as follows:

- Bringing awareness on the significance of threatened biodiversity
- Identifying and documenting the endangered species
- Providing a global index of the decline of biodiversity
- Defining the conservation priorities and guiding the action plans of conservation

IUCN has recognized certain Red List categories of species as extinct, critically endangered, vulnerable, lower risk and rare.

- A taxon is extinct when there is no reasonable doubt that the last individual has died.
- A taxon is critically endangered when it is facing an extremely high risk of extinction in the immediate future.
- A taxon is endangered when it is facing a very high risk of extinction in the near future.
- A taxon is vulnerable when it is facing a high risk of extinction in future.
- A taxon is at lower risk when the size of the population is big and represented by all age groups.
- A taxon with small world populations that are not at present endangered or vulnerable but are at risk.

The critically endangered, endangered and vulnerable categories come under threatened category. The IUCN Red List system was initiated in 1963 and its report of the year 2000 contains assessments of more than 18,000 species of which 11,000 are threatened. The Red Data Books of Indian plants and animals are prepared by Botanical Survey of India and Zoological Survey of India respectively.

The number of plant and animal species of various threatened categories in India are listed below:

Category	*No. of species*	
	Plants	Animals
Critically endangered	44	18
Endangered	113	54
Vulnerable	87	143
Lower risk	73	109

Endangered Flora of India

In India, nearly 244 species of plants are listed under endangered category. Some of them are as follows:

Aconitum deinorrhzum (Monk's hood), *Atropa acuminata* (Deadly night shade), *Colchicum luteum* (Meadow saffron), *Dianthus coschemiricus* (Pink carnation), *Drosera indica* (Sun dew), *Nepenthes khasiana* (Pitcher plant), *Rauwolfia serpentina* (Sarpganth or Sarpgandha), *Pterocarpus santalinus* (Red sandalwood), *Santalum album* (Sandalwood).

Endangered Fauna of India

Almost all the groups of animals are threatened or endangered. Among them mammals, birds and reptiles are the major groups as they are over-exploited and traded illegally. Some of the familiar species of endangered animals are:

(*i*) *Mammals:* Hoolock gibbon (the only ape in India), lion-tailed macaque, Nilgiri langur, Indian pangolin, red panda, Himalayan brown bear, Indian lion, leopard, great Indian one-horned rhinoceros, Indian wild ass, Andaman wild pig, Kashmir stag, brow-antlered deer, Indian bison or gaur, wild buffalo, baleen whales and Gangetic dolphin.

(*ii*) *Birds:* Black eagle, Indian peafowl, black-necked crane, great white crane, Hodgson's frog mouth, great Indian bustard and a variety of hornbills.

(*iii*) *Reptiles:* Green sea turtle, estuarine crocodile, gharial, monitor lizards and Indian python.

(*iv*) *Amphibia:* Viviparous toad and Indian salamander.

(*v*) *Invertebrates:* Coconut crab, dragon flies, butterflies, moths and beetles.

The number of threatened species belonging to different taxonomic groups in India are 459.

4.8 BIOGEOGRAPHICAL ZONES OF INDIA

India has unique climate in the world. Its physical features, the level of land and the mountain ranges influence the climate which in its turn influences the plant and animal life. Though India is a tropical country, it has all ranges of global climate: the tropical, sub-tropical, temperate, desert and Arctic (Tundra) climates. There are two monsoons the south-west and the north-east, besides some torrential rains in between. This variety of climatic conditions permits the existence of diverse group of organisms in different geographic zones.

Biogeography deals with the geographical distribution of plants and animals. The composition of the communities and their distribution depends on the local climatic conditions.

According to a recent classification given by the Wild Life Institute of India, the country has ten bio-geographical zones as shown in Figure 4.2.

(i) **Trans-Himalayas:** This region is characterized by sparse vegetation and rich wild animals as wild sheep, wild goat, snow leopard and migratory black-necked crane.

(ii) **The Himalayas:** These mountains are located at the boundary of many biogeographical regions and are therefore one of the richest areas of the sub-continent in terms of habitat and species diversity. In addition to the species present in trans-Himalayan zone, musk deer, Kashmir stag, and lynx occur in the Himalayas. The forests are dominated by timber trees (*Shorea spp*) pines (*Pinus spp*) and cane (*Salix spp*). In addition to this shrubs and herbs like *Rhododendron, Polygonum* and *Saxifraga* are present.

(iii) **Desert:** North-west of India is the desert region characterized by very

FIGURE 4.2 Biogeographical regions of India.

hot and dry summer with extensive grasslands. The great Indian bustard and wild ass are found here. Its fauna is similar to that found in Palearctic Sahara and the Ethiopian Kalahari desert. The plant species are mostly xerophytic that live on scarce resource of water, such as *Acacia nelotica, Zizyphus spp* and *Calotropis* spp.

(iv) **Semi-arid:** Semi-arid areas are the transitional zones (ecotone) between the desert and the dense forests of western ghats on one side and the Gangetic plains of the lower Himalayas on the other. Lions are found in Gir forests of Gujarat. Vegetation includes *Carissa spinarsum, Acacia leucopholea* and *Zizyphus spp.*

(v) **Western ghats:** This zone extends along the west coast (Malabar) beginning from the Vindhya mountains to the Cape Comorin. This region is characterized by heavy rainfall and is one of the hot spots of India. The vegetation is similar to that of four types of forest: tropical moist ever green forests, mixed deciduous forests, temperate evergreen forests and mangrove forests. Some of the plants found in this zone are bamboos like *Dendrocalamus strictus, Bambusa* *arundinacea* and trees like *Dipterocarpus indicus*. In the Nilgiri, Palani and Annamalai hills there are temperate evergreen forests called *shola forests* with trees such as *Gordonia obtusa* and *Eurya japonica*. These forests provide shelter for elephant, gaur, tiger, Nilgiri langur, the lion-tailed macaque, Nilgiri thar, Malabar civet and giant squirrels.

(vi) **Deccan plateau:** It is a semi-arid region, south of the Vindhyas, lying in the rain shadow of the western ghats. It includes Andhra Pradesh, Tamil Nadu and Karnataka. It is a drier region compared to the western ghats with an annual average rainfall of about 10 cm. It has a central hilly plateau with forests (trees: *Boswellia serrata, Tectona grandis*) and low eastern dry coromandal coast with tropical dry evergreen forests of *Santalum album* (sandalwood), *Cedrella toona, Phyllanthus spp, Euphorbia spp* and *Capparis spp.*

(vii) **Gangetic plain:** This zone extends upto the Himalayan foot hills and comprises Uttar Pradesh, Uttaranchal, Bengal and Bihar. The climate is moderate. Vegetation merges with the adjoining desert zone. Many desert forms of animals like desert cat, desert fox and desert hare live here. The mouth of the Ganges forms a swamp forest, the Sunderbans, and swamp deer is the characteristic species of this habitat. Royal Bengal tiger is forced to lead an amphibious life as the land mass is discontinuous due to the presence of a number of water channels. This is the most fertile zone. Temperature and rainfall, together, are the key factors for the type of vegetation that dwells in this zone. *Rhizophora spp, Acanthus spp,* and *Ceriops spp,* are the plants that occur in the Gangetic delta region.

(viii) **North-East India:** This zone is another rich zone of biodiversity and one of the hot spots of India. This region receives the heaviest rainfall with Cherrapunji as much as more than 1,000 cm per year. The temperature and humidity are very high. Hence, it has dense tropical evergreen forests. It is unique in its flora with several species of orchids, bamboos (*Bambusa pallida*) ferns, conifers (*Pinus khasiya*), insectivores (*Nepenthes spp*) and *Rhododendron spp.* Among the animals, majority of them are enlisted as endangered such as Hoolock gibbon, brow antlered deer, red panda, golden langur, one-horned rhinoceros, elephant and slender loris.

(ix) **Islands:** Islands have a typical defined climate and the species that live there are sensitive to any disturbances. The Lakshdeep islands (in the Arabian Sea) and Andaman and Nicobar islands (in the Bay of Bengal) have evergreen forests. Several species of plants and animals are endemic to these islands. Crab eating macaque, Andaman pig and plant species as *Calophyllum, Mimusops* are worth mentioning among them.

(x) **Coasts:** India has a coastline of 7,000 km. There is an extensive estuarine tract along the coast with mangrove ecosystem such as

Pichavaram near Madras, Ratangiri in Maharashtra, Coringa and Krishna estuaries in Andhra Pradesh and Sunderbans in Bengal. Mangrove ecosystem is found in the intertidal mud flats and estuarine deltas along the tropical and sub-tropical sea coasts. They are dominated by trees. In India the mangrove vegetation includes spp of *Rhizophora, Avicennia* and *Xylocarpus.* Mangrove is a good habitat for a variety of animals too. It is a rearing area for reptiles, crabs and shell fish. Fiddler crab, storks, proboscis monkey and lion-tailed monkey are some of the fauna of mangroves.

4.9 HOT SPOTS OF BIODIVERSITY

Biodiversity is not uniformly observed across the different geographical regions of the earth. Certain regions of the world are megadiversity zones where a very large number of species are found. Most of the megadiversity zones of the world are found near the equator, especially tropical rainforests and coral reefs. Of all the species of the world only 10 to 15 per cent live in North America and Europe. The Malayan peninsula has atleast 8,000 species of flowering plants while Britain, with an area twice as large has only 1,400 species. South America has 2,00,000 species of plants. India accounts for only 2.4 per cent of the land area of the world, but it contributes approximately 8 per cent species to the global diversity.

'Hot spots' are the richest and the most threatened reservoirs of plant and animal life on the earth. The concept of hot spots was developed to designate priority areas for in situ conservation. The key criteria for determining a hot-spot are as follows:

- Number of endemic species i.e. the species which are not found elsewhere
- Degree of threat which is measured in terms of loss of habitat

Twenty-five terrestrial hot spots (Figure 4.3) for conservation of biodiversity have been identified worldwide. These hot spots, together, cover 1.4 per cent of the land area of the earth. Tropical forests appear in fifteen hot spots. As many as sixteen hot spots are in the tropics. About twenty per cent of the human population lives in the hot spot regions.

Among twenty five hot spots of the world, two (Western Ghats and Eastern Himalayas) are found in India and these extend into neighbouring countries also. These areas are rich in flowering plants, reptiles, amphibians, butterflies and some mammals and show high degree of endemism. The eastern Himalayan hot spot extends to the north-eastern India and Bhutan. The temperate forests are found at an altitude of 1780 to 3500 m. Many deep and semi-isolated valleys found in this region are exceptionally rich in endemic plant species. In addition to this, numerous primitive families like Magnoliaceae and Winteraceae with the genera, Magnolia and Betula are found in the eastern Himalaya.

1. Tropical Andes, 2. Mesoamerica, 3. Caribbean, 4. Brazil's Auanue Forests, 5. Choco/Darren/Western Bendor, 6. Brazil's Cerrado, 7. Central Chile, 8. California Floristic Province, 9. Madagascar. 10. Eastern Arc and Coastal Forests of Tanzania/ Kenya 11. West African Forests, 12. Cape Floristic Proviace, 13. Succulent Karog, 14. Mediterranean Basm, 15. Caucasus, 16. Sundland, 17. Wall Area, 18. Philippines, 19. Indo-Burma, 20. South-Central China, 21. Western Ghats/Srilanka, 22. Southwest Australia, 23. New Calenonia, 24. New Zealand, 25. Polynesia/ Micronesia.

FIGURE 4.3 Hot spots of terrestrial biodiversity.

The western ghat region runs parallel to the west coast of Indian peninsula for almost 1600 km. in Maharastra, Karnataka, Tamil Nadu and Kerala. The Agasthyamalai hills, the Silent Valley and the new Amambalam Reserve are the main centres of biodiversity.

4.10 CONSERVATION OF BIODIVERSITY

The extinction and erosion of biodiversity has reached almost its peak, so a time has come to arrest totally the erosion of wild and domesticated biodiversity of the entire planet. The strategy for the conservation of biodiversity is based on two principles; *sustainability* and *equity*. This needs a fundamental shift in planning towards a model, which respects both nature and culture, and empowers local communities to participate in decisions regarding the use of natural resources.

Today in India a multitude of government programmes, run by specialized institutes and agencies are operating to conserve biological resources. To mention a few, there are networks of national parks and sanctuaries, protected forests and biosphere reserves. Other prominent projects aim at saving endangered species, such as Project Tiger, Project Crocodile and the very recent Project Elephant. There is also an expanding network of *ex situ conservation* attempts including botanical and zoological gardens and gene banks. For example, the National Bureau of Plant, Animal and Fish Genetic resources, under the Indian Council of Agricultural Research has undertaken extensive collections and the preservations of germplasm of domesticated species.

A number of international bodies are concerned about the conservation of biodiversity. They aim at protecting species diversity with sustainability and equity through rights of patenting, conservation and use of genetic resources, legal sanctions and protection by law. Some of the organizations are listed below:

ATBI	–	All Taxon Biodiversity Inventory
CITES	–	Convention on Trade in Endangered Species of Wild Fauna and Flora
GEM	–	Global Environmental Markets
IUCN	–	International Union for Conservation of Nature and Natural Resources
MAB	–	UNESCO's Man and Biosphere Programme
UNEP	–	United Nations Environmental Programme
WIPO	–	World Intellectual Property Organization
WWF	–	World Wide Fund for Nature

There are two basic strategies of biodiversity conservation, in situ (on site) and ex situ (off site).

In situ Conservation Strategies

In situ conservation means the conservation of ecosystems, natural habitats and the maintenance and recovery of viable populations of species in their natural surrounding and in the case of domesticated or cultivated species, in the surroundings where they have developed their distinctive properties. (Article 2 of Convention on Biological Diversity). The in situ strategies emphasize protection of total ecosystem. The approach includes protection of a group of typical ecosystems through a network of protected areas.

(i) **Protected areas:** These are the areas of land and/or sea, especially dedicated to the protection and maintenance of biological diversity and of natural and associated cultural resources. These are managed through legal or other effective means. Examples of protected areas are National Parks and Wildlife Sanctuaries. The World Conservation Monitoring Centre has recognized 37,000 protected areas around the world. As of September 2002, India has 581 protected areas (89 National Parks and 492 Wildlife Sanctuaries) covering 4.7 per cent of the land surface as against 10 per cent internationally suggested norm. The Jim Corbett National Park was the first National Park established in India. Some of the important benefits obtained through protected areas are as follows:

- Maintaining viable populations of all native species and subspecies
- Maintaining the number and distribution of communities and habitats and conserving the genetic diversity of all the existing species
- Preventing man created introduction of exotic species
- Making it possible for species/habitats to shift in response to environmental changes

(ii) **Biosphere reserves:** Biosphere reserves are a special category of protected areas of land and/or coastal environments [(as in Figure 4.4)] wherein people are an integral component of the system. These are representative examples of natural biomes and contain unique biological communities. The concept of Biosphere Reserves was launched in 1975 as a part of UNESCO's Man and Biosphere Programme (MBP) dealing with the conservation of ecosystems and the genetic resources contained therein. In India, there are thirteen biosphere reserves and they are also notified as National Parks.

FIGURE 4.4 Biosphere reserves in India.

The main functions of a biosphere reserve are:

(a) *Conservation:* To ensure the conservation of landscapes, ecosystems, species and genetic resources. It also encourages traditional resource use.

(b) *Development:* To promote economic development which is culturally, socially and ecologically sustainable.
(c) *Research and education:* To provide support for research, monitoring, education and information exchange related to local, national and global issues of conservation and development.

(iii) **Sacred forests and sacred water bodies:** A traditional strategy for protecting biodiversity is to declare a forest or a lake or a river a sacred one. The sacred forests vary in dimensions and are protected by tribal communities due to the religious sanctity attributed to them. They represent patches of undisturbed forests without any human impact on them. In India, sacred forests are located in several parts viz. Karnataka, Maharashtra, Kerala and Meghalaya. The sacred forests serve as a reserve for a number of rare, endangered and endemic taxa. Similarly several water bodies in Sikkim, Uttar Pradesh, Maharashtra and Karnataka have been declared sacred by the people leading to the protection of aquatic flora and fauna.

The conservation efforts towards plant species have not been given adequate attention particularly of those which are of potential economic and scientific value. Among the animals, the approach to conservation has been restricted to large mammals as they are at the top of the food chain. Instead of this, a holistic approach would save much of the biological wealth of the country. The thrust areas in the in situ conservation strategies are as follows:

(i) Relations between plant and animal species
(ii) Quantitative assessment of the conservation status of the species
(iii) Multiplication and restoration of endangered, rare and endemic species using biotechnology
(iv) Identification of critical index species and their sensitive parameters
(v) Assessment of the impact of exotic species on the ecosystem
(vi) The possible/predictable climate change and its impact on biodiversity
(vii) Hydrological changes including surface run off and percolation in the protected areas
(viii) Primary production and cycling of nutrients in the soil
(ix) Development of methodologies for classification of microhabitats
(x) Ecological restoration of degraded micro and macrohabitats
(xi) Study of remotely sensed data (satellite maps) of all protected areas

Ex situ Conservation Strategies

In the year 1992, an earth summit was conducted in Rio de Janeiro wherein about 150 nations signed the convention on biological diversity. The convention became International Environmental Law in the following year. The preamble of the convention covers issues related to the sovereign rights of the states over their biological resources, biodiversity and its conservation. One of the notifications of the preamble says that "ex situ" measures in the country of origin have an important role to play in conservation. In the Article 2 ex situ conservation is defined as the conservation of components of biological diversity outside their natural habitat.

The methods of ex situ conservation are expensive and need a careful planning and systematic approach. These strategies include the maintenance of botanical gardens, zoos, aquaria, gene banks, seed banks, use of biotechnology and DNA preservation.

(i) **Botanical gardens and zoos:** Conservation of species in botanical gardens and arboreta is already in practice. There are more than 1,500 botanical gardens and arboreta (a plot of land on which trees or shrubs are grown for study or display) in the world with more than 80,000 species. Many of these have now seed banks, tissue culture units and other ex situ technologies. Similarly there are more than 800 professionally managed zoos around the world with about 3,000 species of mammals, birds, reptiles and amphibians. Many of them have well-developed captive breeding programmes for endangered species. Aquaria maintain approximately 5,800 species of fish and most of them are collected from the wild.

(ii) **Biotechnological methods:** Biotechnology has provided many new conservation tools in value areas of agriculture, animal husbandry, fisheries, forestry and medicine. Creation of transgenic crops which can withstand the pests and environmental stress is a landmark in the field of applied Biotechnology. However, an indiscriminate introduction of such genetically modified organisms poses a threat to biodiversity. In fact, uniformity (or homogeneity) in life forms accelerates the loss of biodiversity. Hence a rationale should be followed with vigilance in applying the tools of biotechnology for conservation practices.

(iii) **Gene banks:** Gene banking provides a method of conservation of diverse genetic resources, particularly of threatened species and those seeds which are not viable for longer periods under natural conditions.

Seeds of plants can stay alive as long as there is food reserve and water content inside. Several vital chemical reactions will be occurring inside the seeds to keep them alive. The rate of these reactions are directly dependent on the ambient temperature and the water content of the seed. The higher the temperature, the faster the reactions that take place, and lesser the period for the food reserves to last. Based on this fact, the temperature of the surroundings and water content of the seed are reduced to prolong the life of the seeds. A simple technique employed in the preservation of seeds is cryogenic preservation, which involves suspended animation at sub-freezing temperatures and low moisture content. The seeds are desiccated and maintained at –15°C in simple chest freezers.

The United States National Seed Storage Laboratory (NSSL), Vavilov Institute in Leningrad, Izmir Centre in Turkey, International Rice Research Institute (IRRI) in the Philippines, Royal Botanical Gardens at Kew and Seed Bank at University of California, Irvine, are some of the pioneer gene banks at global level. India has a number of germplasm centres where collections (accessions) of germplasm of varieties of plants are stored. To mention one among them is the Rice germplasm centre in Indira Gandhi Agricultural University of Raipur (Chhattisgarh) where 19,000 varieties of rice germplasm are collected and stored.

(iv) **Conservation of DNA:** An emerging and promising technique in preserving biodiversity is isolation and conservation of DNA. It can be used for endangered or even extinct species by taking samples of material from hair, bones, and herbarium specimens of the target species.

In addition to the above-mentioned strategies, restoration and rehabilitation of species, populations and ecosystems can be achieved through relocation, reintroduction, artificial incubation, artificial insemination, embryo manipulations and propagation.

Problems in Conservation

1. One basic problem is very little understanding of what is to be conserved, especially with regard to complex natural ecosystems like tropical rain forests. Conservation attempts are patchy, focussed on only few visible, big animals and forests. But an ideal method is to conserve all ecosystems with entire biota.
2. Financial resources to protect and manage ecosystems are inadequate. Staff and forest guards work under dangerous conditions whereas the poachers are equipped with sophisticated weapons.
3. Alienation of people from their natural resources bring in resentment among local people and no protected area can survive under hostile social environment. Wildlife needs to be protected with the people rather than from the people.

Legal Coverage

Attempts have been made to give biodiversity conservation some legal sanction. As many as 20 Acts are relevant to biodiversity. Some of the important Acts are as follows:

(i) Wild Life (protection) Act 1972
(ii) Forest (conservation) Act 1980
(iii) Wild Birds and Animals Protection Act 1887
(iv) Indian Forest Act XVI 1927

In fact, India is the pioneer country in bringing in the worldwide awareness on the need for conservation of species. British Natural History and Society (BNHS), Bombay and Central and State Boards of Wildlife are putting in commendable efforts towards conservation of biodiversity.

Indian Scenario

Though several acts and laws are in vogue, implementation of the same is not completely observed. There are plenty of illegal activities going on against the rules of conservation of wildlife.

Sunderbans is a vast estuarine mangrove swamp forest protected since 1878. But in the post-independence years, wildlife has disappeared rapidly due to poaching, deforestation and resettlement of displaced human populations. In addition to this, the hunting places of Royal Bengal tiger were disturbed by man to the extent that it turned into a man-eater.

On the Andaman and Nicobar Islands, threat to wildlife has been increasing due to immigration of people, sand mining from beaches, commercial exploitation of timber, and disturbance caused by the highway that is passing through the Jarawa reserve. In this regard a joint petition was filed in 1998 by the Society for Andaman and Nicobar Ecology (SANE), Port Blair, and BNHS, Bombay. The Supreme Court gave the orders to this effect putting ban on such human activities.

Intellectual Property Rights (IPR)

Plant Variety Protection and Farmers Rights Act and the Biological Diversity Act 2002 create intellectual property rights and access to biological resources. These IPR legislations are taking away resources from the common to a private regime. In addition, the International Treaty on Plant Genetic Resources for Food and Agriculture (ITPGRFA) allows companies to access common resources and permits private property regimes. The Biological Diversity Act 2002 provides for a regulatory system to get permission and access to biological resources. This has to be routed through the National Biodiversity Authority (NBA) which has the right to deny access. However, it is not clear, whether this is practical or not. For example, if a country has endorsed the ITPGRFA, then the country is obliged to grant access to any of the crops/resources. The best known example is the battle over the patent rights on neem tree. Another similar issue on the ownership of germplasm is the claim of Syngenta, India (an affliate of Syngenta, AG, Switzerland) on the rice varieties in Chhattisgarh. The Indira Gandhi Agricultural University of Raipur has 1,90,000 collections of rice germplasm (varieties of rice). Syngenta, India has negotiated a tie-up with the Univeristy for a rice research programme using this germplasm. But the University authorities and the environment activists opposed to this proposal. However, Syngenta has cited ITPGRFA claiming their access to the natural resources in India. In this bargain majority of the developing countries are losing their rights on their own resources.

Recommendations

(i) The responsibility of protecting the natural resources of the country must remain with its people. Farmers will be the losers in a centralized patent regime that gives preference to monopolies.

(ii) The Acts and laws should help the farmers in protecting the sovereignty and access to the resources.

(iii) The germplasm belongs to the community and a community or collective ownership must be established instead of a tieup with multinationals.

Success Stories

Every dark cloud has a silver lining. In the previous sections, threats to wildlife, deforestation, disturbances in ecosystem and other perils have been mentioned in extenso. The whole picture may be frightening and fatal. However, there are a few cheerful and optimistic outlets too.

India has done commendable job to save its natural resources. Following the international guidelines 89 national parks and 492 wildlife sanctuaries have been set up and 13 biosphere reserves have been marked off. One of the most spectacular successes in wildlife management in India has been the Project Tiger launched in 1973. By 1987 the tiger population has increased from 273 in 1972 to 1229 (in the 18 tiger reserves). The objective of the project was not confined to tigers but extended to preserving the total ecosystems. Other species as the swamp deer, the elephant, the rhinoceros and the wild buffalo also multiplied in their number as their habitat is preserved.

Another fruitful programme on wildlife management is that of the Asiatic lion (Panthera leo persica). Sometime back lion was on the verge of extinction but the Gir forests in Gujarat were protected and successful conservation strategies have revived the number. Similarly crocodiles have been saved by legislation. Both mugger and gharial have been declared protected and a project was started in 1975 with the assistance of United Nation (UN). There are two model sancturies in Orissa and Andhra Pradesh and crocodile farming is actively going on in Tamil Nadu, Uttar Pradesh, Rajasthan, Gujarat and Kerala. In addition to this, India has managed to revive the population of two birds, cheer pheasant and the great Indian bustard. The protection of Silent Valley of Kerala has brought many plant and animal species to light which were not recorded earlier.

Restoration of Chilka lagoon (Orissa) along the east coast of India can be quoted for a triumphant protection of aquatic ecosystem. It was observed that certain indicator species like Irrawady dolphin, sea grass and fish have declined in their numbers. A combined action of government agencies and local community saved the ecosystem. In their action plan, hyacinth and other weeds were removed from the lagoon and a new mouth for water exit to sea was made which proved to be successful. As a result, a healthy sign of growth was observed in the index species and the fishery landings increased gradually from the year 2000 to 2003.

No single agency or government can possibly achieve the task of conserving biodiversity, particularly in a complex society as India's. The total participation of people at all levels (non-government organizations and civilians) for managing the resources of the country should be considered a fundamental right and prerequisite for successful conservation.

SUMMARY

Biodiversity refers to the variety and variability among living organisms and their surrounding nature of ecosystem influenced by several factors like climate, resources, disturbances. Each and every component has its unique value as food, medicine, wealth, recreation, culture and other benefits to mankind. Even small things are more important for the stability of an ecosystem. As a part of evolutionary changes with time, arrival of new species and elimination of older ones appear to be a normal process. But human activities in the past few decades are increasing this rate of extinction of species. The richest reservoirs of plant and animal wealth (hot spots) are more threatened due to factors like loss of habitat, deforestation, poaching, pollution, pests. Some species are more vulnerable due to some distinctive features like large body size, low reproductivity rate, scarcity of nutrient materials, poor living conditions. One aspect of conservation is to establish national parks, wildlife sanctuaries, gene banks, biosphere reserves and undertaking prominent schemes like project tiger. The other method is to protect the ecosystem i.e. the surrounding environment. Wildlife and Indian forest acts accord legal measures but the responsibility of protection remains with the people of India.

KEYWORDS

Conservation strategies, endangered species, hot spots, national parks, plants and animal wealth of India, significance of diversity, threats, wildlife protection and forest conservation acts, wildlife sanctuaries.

SOLVED QUESTIONS

1. **State the main reasons for variability among living organisms.**

   The main reasons for variability are:

   (i) Geographical regions with variations in temperature, rainfall, soil conditions from poles to equator

   (ii) Ecosystem variations such as forests, marine and fresh water bodies

   (iii) Living conditions, life cycles, living styles, inter-relationships among various organisms

   (iv) Availability and utilization of food and energy sources in the system

   (v) Natural processes for selection, survival, succession and evolution

2. **What are the benefits from different organisms and diverse conditions?**

   The benefits are:

   (i) Stability in climate, water, soil etc. in the biosphere
   (ii) Food, fodder, fuel, fibre, medicines and other material needs
   (iii) Biogeochemical cycles in nature including carbon fixation, nutrient recycling and decomposition of wastes
   (iv) Recreation, continuance of life processes

3. **List out the human activities that results in loss of biodiversity.**

   Human activities that results in the loss of biodiversity are:

   (i) Human activities disturb the natural environment and influence the harmonious living conditions for biota
   (ii) Population explosion demanding more resources
   (iii) Urbanization, consumerism and market needs
   (iv) Disposal of wastes leading to depletion and degradation of environmental resources
   (v) Industrialization, river valley projects, mining and other activities
   (vi) Destruction of forests, overgrazing, soil erosion, desertification.
   (vii) Exploitation, illegal trade activities, new technologies and materials
   (viii) Tourism and recreational activities.
   (ix) Elimination of even some species damage the total ecosystem
   (x) Extinction is natural in an ever changing system
   (xi) Global climatic changes

4. **How does biotechnology help in conserving biodiversity?**

   Biotechnology conserves biodiversity by

   (i) Selection of crop varieties to suit the local conditions
   (ii) Identifying genes for resisting pesticides and transferring to plants
   (iii) Discovering new bacterium for oil destruction and for biodegradation of plastics
   (iv) Establishing international gene and seed banks to help farmers
   (v) Developing tissue culture methods and artificial insemination techniques for conservation of endangered, plants and animal species

5. **List some heritage sites for conservation of biodiversity.**

   Sambhar and Keoladeo in Rajasthan, Chilka in Orissa, Harika in Punjab, Sunderbans in West Bangal, Nanda devi, Uttar Pradesh, Manas in Assam, Nilgiri in Tamil Nadu and Nokrek in Meghalaya are heritage sites

6. **Write a note on the locations of national parks and wildlife sanctuaries in India.**

National parks	*Wildlife sanctuaries*
Dachigam (Jammu & Kashmir)	Sultanpur (Haryana)
Kaziranga (Assam)	Jalda Para (West Bengal)
Gir (Gujarat)	Hazaribagh (Bihar)
Kanha (Madhya Pradesh)	Ghana (Rajasthan)
Corbett, Dudwa (Uttar Pradesh)	Abohar (Punjab)
Sariska, Ranthambore (Rajasthan)	Nalsarovar (Gujarat)
Bandipur (Karnataka)	Mudumalai (Tamil Nadu)
Periyar (Kerala)	Vedanthagal (Tamil Nadu)

7. Name few critically endangered species.

Black and northern white Rhinos (Africa), Giant panda (China), Bamboo lemur (Madagasca) snow leopard (China and Russia), orantung and Sumantran tiger (Siberia and China), Mountain Gorialla (Congo and Uganda), Salmon crested cockatoo (Indonesia).

In India,

Reptiles	–	gharial, python, green sea turtle
Birds	–	bustard, crane, peacock
Mammals	–	wolf, red fox, sloth bear, red panda, striped hyena, tiger lion
Primates	–	nilgiri langur, lion-tailed macaque, capped monkey

REVIEW QUESTIONS

1. Define genetic diversity and explain the processes by which it arises in a population.

2. Species richness is an index of species diversity. Explain.

3. What is the effect of habitat loss on biodiversity?

4. What are the characteristics that make a species susceptible to extinction?

5. What are biosphere reserves? How do they help in the conservation of species?

6. Why should we conserve biodiversity? Give reasons.

7. India is a megadiversity nation. Justify.

8. Anthropocentric activities, to a large extent are responsible for the loss of biodiversity. Explain.

9. Give an account of the endangered flora and fauna of India.

10. What are the major in situ strategies of conservation of biodiversity?

11. Describe biodiversity with reference to the biogeographical regions of India.

OBJECTIVE TYPE QUESTIONS

Fill in the blanks with suitable words:

1. The species that are confirmed to a particular geographical area are called ______________ species.
2. ________________ are the richest and the most threatened reservoirs of plant and animal life on earth.
3. The two hot spots of biodiversity in India are _______________ and ______________.
4. Which of the following is the number of chromosomes ($2n$) present in man?

   (a) 48 (b) 46 (c) 32 (d) 28
5. Red list catalogue of taxa includes those species that are:

   (a) harmful to human beings
   (b) facing the risk of extinction
   (c) responsible for pollution of the environment
6. Expand the following:

   DNA, IUCN, MAB, UNEP
7. Match the following:

1. Histone	Anti-cancer drug	( )
2. Polymorphism	Endangered Indian ape	( )
3. Vimblastine	Loss of wildlife	( )
4. Hodgson's frog mouth	Shades of skin colour	( )
5. Poaching	Protien	( )
6. *Santalum album*	Endemic bird in India	( )
7. *Hoolock gibbon*	Sandal wood	( )

CHAPTER 5

COMMUNITY HEALTH

TOPICS COVERED

- Population Studies
- Disease Transmission
- Environment and Health
- Environmental Sanitation Measures
- HIV and AIDS

5.1 POPULATION STUDIES

Demography is the study of population growth, composition and distribution of population on the land. The following facts are observed from the statistical records.

- One out of every seven people in the world is an Indian.
- Present population of India is around 1000 millions, second to China.
- 75 per cent of India's population live in 6 lakh villages.
- Population density in India is more than 300 persons per sq. km of land, and is mostly concentrated in cities and urban slums. On the other hand, world's average population density is around 50 persons per sq. km.
- Indian population is three times more than that of Western Europe whereas both the countries have the same land area (i.e. 3.3 million sq. km).
- China's population is 20 per cent more than that of India. USA has one third of India's population. However China (and also USA) has three times more land area than India.

Present population figures and land areas of some countries are shown in Figure 5.1. More people live at places where resources, employment,

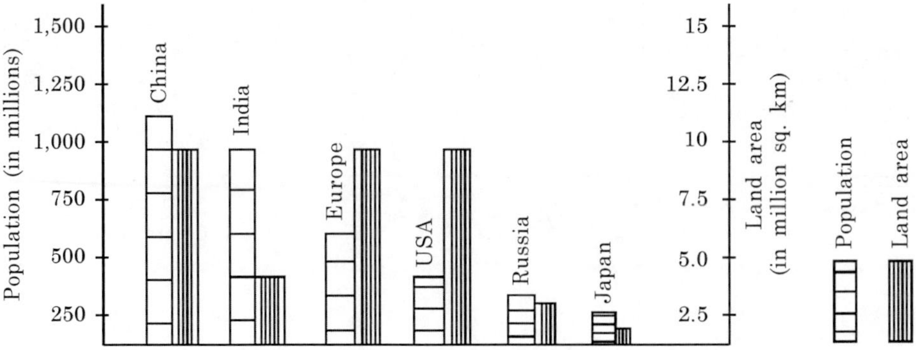

FIGURE 5.1 Population figures.

opportunities and other facilities are available. This leads to over utilization and exploitation of resources resulting in their depletion and degradation. Developed countries have less population and better standards of living whereas most populous countries are struggling even when resources are available in plenty. 25 per cent of the population occupy 45 per cent land area, utilize 65 per cent of energy and earn 85 per cent of world's income in the developed countries. The greatest obstacle to the socio-economic advancement of under-developing world is its rapid increase in the population every day. For example, population in India is increasing at a rate of above 2 per cent per year and doubling in about 45 years of time. In the last 50 years, agricultural land has decreased from 0.5 ha/head to 0.15 ha/head. Also life expectancy is improving over the years due to welfare measures adopted in independent India. When people in the age group of 0–15 years and above 60 years, are more in number, dependency increases and productivity recedes.

Progress of a nation depends on a number of factors such as rate of population growth, population density, female to male ratio, literacy rates, age structure, family size, mortality rates, per capita income, standards of living and social welfare measures implemented.

Population growth and population density exert greater demand on land, water, food, energy and other resources. More living space, more education, health, communication, industries and other basic facilities are needed. Construction industry converts agricultural lands and water bodies into residential units. Overgrazing, deforestation, mining and other activities cause land degradation. Use of chemical fertilizers and pesticides for higher yield of agricultural product causes soil infertility. Change in life styles, consumerism, market trends, human actions lead to ecological imbalances. Solid, liquid and gaseous wastes pollute all natural resources resulting in unsustainable environment. Scarcity and exploitation of resources lead to unrest among the group of individuals in a society. As long as populations continue to increase in alarming numbers, any extent of development of a nation appears to be negligibly small. Hence, the major problem lies in controlling the increasing growth rate of population as death rates at all levels have been reduced drastically due to the human welfare programmes in the country.

Figure 5.2 shows a typical trend in population growth at a place. Initially, people take time to adjust and live in a new environment. Growth of population will be exponential in nature, when plenty of resources are available. But they cannot go on increasing forever as resources are likely to be shared by lots of people. Finally a stage of saturation will be reached in any community depending upon the carrying capacity of the environmental resources.

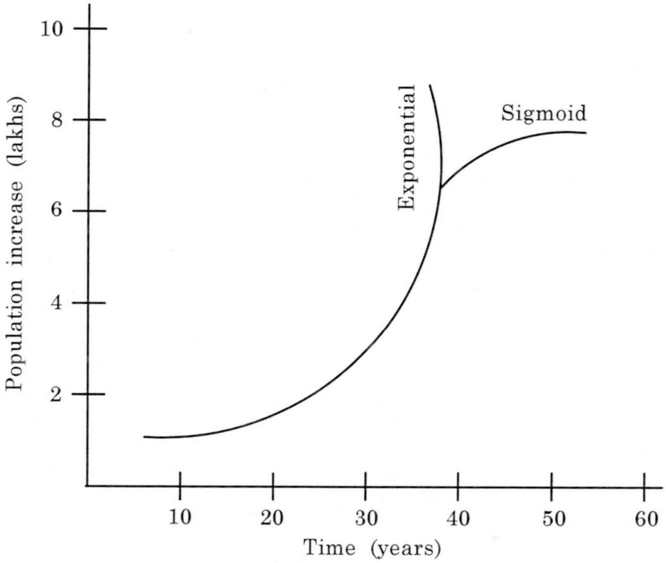

FIGURE 5.2 Population growth patterns.

Average life expectancy is increasing due to the advancement in medical and social infrastructure. Figure 5.3 reveals that average age at the time of death was 45 years in 1961, whereas it is 60 years in 2001. Similarly, mortality rates in a town have reduced (Figure 5.4) due to the implementation of potable drinking water and sewage treatment systems.

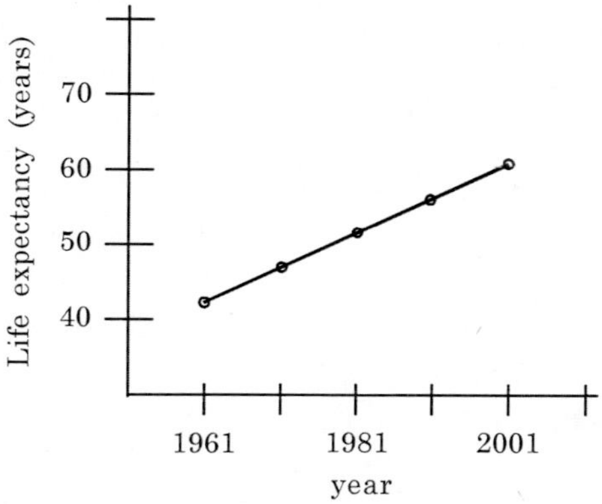

FIGURE 5.3 Life expectancy rates.

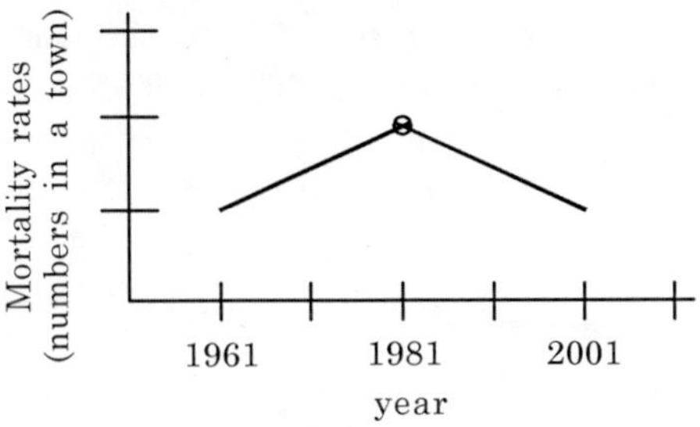

FIGURE 5.4 **Effects of sanitation measures on mortality rates.**

Population Welfare

(i) **Awareness:** Information regarding the merits of a small family norm, methods of prevention and control of birth and availability of welfare facilities within the reach of a common citizen, can easily be communicated through any media source like newspaper, radio, television and cinema. Knowledge is information, awareness is education and wisdom lies in decision making. Age at marriage (preferably 21 years for females and 25 years for males), spacing of three to five years between the first and second child (by using contraceptive pills or condoms) and regulating the family size with one or two children (by sterilization) are desirable practices. There can be incentives for adopting family planning measures and penalties otherwise.

The place where literacy rates, standards of living and per capita income levels are highest, birth rates in that society are less. Infant mortality, male child preference, old age care, social insecurity, children earning and supporting parents, may be some economic reasons for developing large families. Some people may think that more people mean more production of goods or services and thus economic prosperity. Employment potential also increases migration trend towards urbanization.

In Kerala and southern parts of India, literacy rate is very high and the fertility rate (number of children in a family) is low in comparison with other states like Bihar and Rajasthan. Similarly development of a country is related to the growth rate of population.

(ii) **Welfare measures:** It is the primary duty of the Government to provide the basic needs for all the sections of the society.

- *Health for all:* General hospitals for men, women and children at revenue, mandal and district levels, specialist hospitals for maternity and child health, for infectious and communicable diseases and super speciality hospitals for mental health, heart and other diseases are needed.
- *Education for all:* Schools, colleges, professional and vocational

institutions must be available with quality education for the needy. Education should be made compulsory upto an age of 15 years, in practice. Incentives may be given to females and rural folk upto the college education. Part time and distance education may also be encouraged.

- *Communications:* Highways, railways, telephones are essential for quick transport of goods, business activities and income generation.
- Food, shelter, clothing are essential for all human beings. Employment opportunities must be generated so that per capita income increases.

Welfare measures are meant for creating the social well-being and happy environment.

Women and Child Welfare

Indian women stood for a significant tradition and culture since vedic age. However, they are the suppressed, neglected and harassed due to gender discrimination. Prenatal diagnostic tests for determination of sex of a preborn child are prohibited, but foeticide and infanticide of a girl child is happening due to preference for a son. Girls are discontinued from schools, to work in domestic sector. They suffer from malnutrition. They are underpaid and cannot exercise equal rights with men. Anti-dowry act remains on paper only. Women suffered all sorts of abuse by men. Female to male ratio is only 927/1000 and even less in some parts of the country. Now, the government has realized that progress cannot be achieved without adopting women welfare programmes. So, special care (reservations/incentives) has been taken to provide education, health, job and other facilities to women. When they are economically standing on an equal pedestal with men, they can decide for themselves. They can take care of their families and society. That has been a well observed fact in the developed countries.

Proper care is not given to mother during pregnancy and to the new born child in the past due to poverty, illiteracy, ignorance, malnutrition and lack of medical facilities. Infantile mortality rate has been considerably reduced due to various immunization programmes for the children against diseases like polio, BCG, measles, diptheria, tetanus etc. Health programmes are continued in the schools also. Child guidance clinics are available for street children. Other mentally retarded and behavioural problems of children are also taken care.

Nation's welfare starts from an individual's home. An educated female is expected to take care of the family, the future citizens and contribute to the economy of the nation.

5.2 DISEASE TRANSMISSION

Louis Pasteur and Koch postulated germ theory of disease and principles of bacteriology. Edward Jenner discovered vaccination against small pox and later vaccines have been developed for preventing a number of diseases. Causative agents (microbes) of typhoid, tuberculosis, cholera, diptheria and various other diseases have been identified. By 1900, people realized that clean water and clean surroundings block major channels of transmission of disease from person to person.

A person with infection can be treated. It involves money, suffering and loss of production value. When number of persons are affected, it is safe and economical to prevent the transmission of disease than curing individuals. One cannot be kept in isolation or quarantine for long period due to number of reasons. Some times it appears that a disease is cured but still the person is in a position to communicate the disease. It is practically not possible to control the activities of all persons suffering from diseases. Similarly protection cannot be given to each and every person by adopting a number of immunization programmes starting from the new born to the aged, under ideal conditions. The practical difficulties are—a major percentage of population is poor, uneducated and scattered beyond easy reach. So the best solution lies in controlling the environmental factors only.

Diseases are classified as epidemic (spreading at alarming rates, like cholera), endemic (prevalent only in certain local area, like filaria) and sporadic (scattered in different parts, like polio).

Most of the diseases are likely to be transmitted through environmental route (Table 5.1) from person to person as they are living in the same environment and using the same resources. Diseases are classified as follows, depending upon the transmission mode:

(a) Water borne (cholera, typhoid, jaundice, polio, dysentery, diarrohea)
(b) Air borne (tuberculosis, influenza, bronchitis)
(c) Food borne (enteritis, brucellosis, salmonellosis, trichinosis)
(d) Insect borne (malaria, filaria, yellow fever, gastroenteritis, amoebiasis)

Table 5.1 Disease transmission

	Environmental components	
Person (reservoir of disease causing germs)	Water, air, soil, food, milk, meat, insects, fly, mosquito bacteria, virus (HIV) toxic chemicals, inanimate objects, allergens	Person (susceptible host)

Flies mechanically transfer the disease causative organisms from filth to food items. Anopheles mosquito causes malaria while Culex variety causes filaria. Mosquitoes breed in stagnant water and feed on shrubs. Flies live in unhygienic surroundings. Infected food, milk, poultry and meat products may cause various intestinal disorders resulting in water pollution. A number of harmful chemicals may be released into the atmosphere creating toxic conditions in the living environment.

Diseases may spread from person to person in different ways.

- By direct contact (sexually transmitted diseases), droplets (tuberculosis, diphtheria), soil (hook worms), skin and mucosa (rabies) and from mother to child.
- By indirect contact, vectors (insects like fly, mosquito, and rats), food, milk and eatables, media (water, air) and inanimate objects (towels, bed linens, containers, railings).
- By the release of chemicals into the environment (agricultural, industrial processes).

Sometimes the causes of a disease may be multifactorial. In addition to the communicable diseases that spread through the environment, care has to be taken to safeguard the health from diseases due to malnutrition (vitamin, mineral, protein and iodine deficiency), fatal diseases (cancer, diabetes, heart, kidney and mental problems) and drugs (including alcohol, smoking). Government has established various schemes to protect the health on a permanent basis, such as National malaria eradication programme, Filaria control, Leprosy eradication, Iodine deficiency, Aids control, T.B. control, Maternity and Child health programmes, and Universal immunization programmes. World Health Organization has defined **health** as 'a state of complete physical, mental and social well being and ability to lead a complete productive life'. Unless positive, quality and balanced state of health (including improvement of socio-economic considerations) is developed in the total environment, unrest leads to crimes and miserable life styles in the society.

Occupational Health

People work with machines and chemicals in various factories for major part of a day. They are exposed to physical stress, temperature, humidity, noise radiation, vibrations, glare, infections and toxic chemicals (such as Pb, Hg, As, Cd, CN, Co etc.) in processing units. Leakages and mechanical failures lead to accidents and injuries to workers. An occupational health plan should include the following steps:

(a) Regular health check up for all workers
(b) Adoption of precautionary procedures without fail
(c) Use of personal protective measures
(d) Periodical maintenance of machines and good house keeping
(e) Storage and handling all dangerous chemicals properly
(f) Training to the concerned workers regarding safety measures
(g) Proper disposal of solid, liquid and gaseous effluents.

5.3 ENVIRONMENT AND HEALTH

Water, air, soil and surroundings constitute our physical environment. Plants, animals, insects and microbes form the biological component. Social customs, habits, lifestyles, per capita income, literacy rate and other living conditions also affect the human beings in the total environment. Geographically, climatic conditions (temperature, humidity, rainfall etc.) at a place are fixed by parameters such as nearness to the sea, altitude of a place, latitude and longitude. They also decide the nature of the soil, availability of water and wind movement. In turn, people select their housing, clothing, food and other life patterns. Modernization has led to industrialization, urbanization, mining, construction of dams, deforestation, use of chemical fertilizers and pesticides and a lot of material comforts. All these things created different types of pollution as there was no control on the disposal of solid, liquid and gaseous wastes in the human environment.

Water Pollution

Impurities present beyond safe permissible levels cause harmful effects. They may impart unpleasant taste and odour, turbidity, colour, acidity, alkalinity, hardness to the water and remove the dissolved oxygen content from the drinking water. Sulphates, chlorides, nitrates, fluorides and toxic chemicals present in the water may be injurious to human health. Contaminated water causes water borne diseases. Sometimes, chemical present in the natural rock formations may also dissolve or mixes with the water. Domestic wastes and various industrial effluents may be disposed into the water sources without complete purification. All these impurities may also affect fish and other aquatic life in addition to the human population.

Soil Pollution

Soil fertility is affected by the excessive use of chemicals in the agricultural operations. Due to population explosion, more land is converted into non-agricultural purposes. Even lakes and ponds are reducing in their size. Impurities get deposited in the soil pores causing soil pollution. Crops that are grown on these lands carry pollutants in higher concentrations. Soil once contaminated may become sick and ultimately may not be recovered.

Air Pollution

CO, SO_2, NO_2, HC, particulate matter and a number of chemicals are released into the air from various industrial practices. These chemicals further react within themselves and with the environment, causing serious damage to the living atmosphere. Carbon monoxide reduces the oxygen level in the blood, carbon dioxide causes asphyxiation, and other chemicals released into the atmosphere in high concentrations affect the respiratory and vital organs of human being.

Hearing to loud noise adversely affects the ability to concentrate on the work leading to irritation, headache, tension and hearing loss. Marine water is polluted with toxic chemicals or floating oils endangering the ecosystem. Any component in the total system if present in excess may be injurious to some living organism in the environment.

5.4 ENVIRONMENTAL SANITATION MEASURES

Sanitation measures are intended only to improve the quality of the surrounding and control the environmental factor which causes transmission of disease in the society.

They may be listed as follows:

- Good housing facilities and suitable living conditions
- Supplying adequate quality and quantity of drinking water
- Collecting, treating and disposing safely all types of waste from the domestic environment
- Not to allow suitable living atmosphere for insects, microbes and others near to residential unit
- Protecting eatables from contamination at all stages before human consumption
- Careful use of chemicals that are toxic in nature. Kill microbes and pests
- Taking adequate care when people meet in large numbers any time
- Removing slums and overcrowded spots with inadequate facilities
- Periodically adopting universal immunization practices
- Creating awareness in the society with regards to sanitation measures
- Cleanliness in a school, hospital, slaughter house or any place of living, as cleanliness is next to Godliness

It has been observed that at places where all the above facilities are provided, morbidity and mortality rates have decreased. Community welfare can only be achieved with the cooperation of various departments and the public.

5.5 HIV AND AIDS

Human immuno deficiency virus destroys the body's defense mechanism against any disease and leads to Acquired Immuno Deficiency Syndrome (AIDS). First case was reported 20 years back and virus has been recognized. Out of the four crore patients in the world, thirty lakh Indians are affected and Andhra Pradesh alone has listed about three lakh patients. In the initial incubation period, symptoms of AIDS are loss of weight, fever, diarrohea, cough, rashes, etc. with no solution. **Elisa test** may be used to diagnose the disease, **western blot test** for confirmation, **CD 4** test to know the extent of virus spread and **viral load** test for serious patients under treatment.

HIV does not mean AIDS. Slowly defense mechanism fails and the case may take even 5 to 10 years depending upon the patient's health condition, living habits and life style. The disease spreads through direct transfer of body fluids (semen, blood) containing the virus from an infected person into the blood of another person i.e.

(a) mainly through sexual contact with multiple partners
(b) blood transfusion, through unknown donor's uncertified blood or through wounds and exposures on body surface

(c) injection needles contaminated before use and razors used by many people
(d) mother to the new born child (a very serious concern)

There is no medicine, cure or vaccine at present. Hence care is only to be taken to prevent the disease. That is why AIDS control society, and a number of voluntary organizations are educating the people and creating awareness. So, preventive steps are as follows:

- Avoid extramarital relationships.
- Use precautions (as condom).
- Use disposable syringes.
- Accept blood from approved blood donors only.
- Do not transfer to the partner and new born child.
- Observe moral values and responsible behaviour.

There is no cause for panic as one can live for 5 to 10 years safely with HIV. Friends and family members can support an AIDS patient as the disease will not spread through cough, mosquito, glassware, telephones, toilets etc. Positive approach is very beneficial even when other things fail.

SUMMARY

India's population exceeded 1 billion mark on 1-3-1991, out of which 75 per cent live in villages whereas the population density in cities and urban slums is very high. India has less than 3 per cent of world resources where 15 per cent of the world's population live. This leads to exploitation of resources resulting in chaotic conditions. Population welfare measures are, increasing the literacy rates, standards of living, women and child care, sanitation measures, prevention of transmission of diseases and overall social health conditions. Water, air, soil and surroundings constitute our physical environment. Plants animals and microbes form the biological component. Modernization has created different types of pollution and impurities beyond permissible levels are causing severe harmful effects to the human population.

Community welfare can be achieved easily by providing safe drinking water, adequate drainage facilities and clean surroundings.

Improved living conditions means a happy and prosperous nation.

KEYWORDS

AIDS, demography, disease transmission, environmental sanitation measures, growth and welfare, occupational hazards, population density, women and child care.

SOLVED QUESTIONS

1. **Briefly explain the patterns of increase in population figures and the factors affecting.**

   In a new habitat, food and other resources are available in plenty and harmful interactions with other organisms are less. Though the rate of increase is slow in the beginning, it will increase exponentially (if unchecked) later due to lot of facilities around for growth. More population means more material requirements, overcrowding and unemployment. Population growth maintains a stable equilibrium under optimum utilization of natural resources till it exceeds the maximum carrying capacity of the environment.

   According to Malthus, famine, disease or a natural disaster reduces the population when it is beyond the sustaining capacity of the environment.

   Industrial and agricultural revolutions, scientific progress, welfare measures and other developmental activities influence population numbers.

   Karl Marx believes that social justice solves the population problem as people always sustain, given the equitable share for their fruits of labour.

   Life expectancy increased because of health care, nutrition, education, sanitation and other measures. Improved living conditions also reduce population growth rate. Population densities are not uniform as when more facilities are available at a place, people migrate.

2. **Mention the steps required to check the population explosion.**

   Population growth provides necessary work force, family security and pleasure for the individual to some extent, but finally uncontrolled explosions leads to overcrowding, poverty, crime and pollution all-round.

   Socio-economic development including family welfare measures are aimed at controlling the population growth rate. Education, job opportunities, property rights and health care for women, provide the advantage of decision making in family matters which leads to have children by choice.

   Population densities are high at some places. Steps are to be taken to develop new locations for the uniform distribution of population to the extent possible.

   Lower income groups should be encouraged to follow the required family planning methods as the data over thirty year period indicates more children in that class of people.

Births, marriages and deaths should be compulsorily registered with government authorities.

3. **Outline the measures to be taken to control flies and mosquitoes in domestic environment.**

Fly control measures:

- Eliminate the breeding places (decomposing wastes including garbage).
- Store the food articles in proper containers at proper places.
- Keep the surroundings clean and dry.
- Use insecticides like gammaxene, pyrethrum, malathion to kill flies.

Mosquito control measures:

- Remove small shrubs and stagnant water pools in the domestic surroundings.
- Fit the door and window openings with fine screens to prevent the entry of all vectors.
- Use insecticides like lindane, pyrethrum, dialdrin, parisgreen, to kill mosquitoes in larva and adult stages.
- Use bed nets to prevent mosquito bites during nights.

4. **List out the various harmful chemicals released into the environment.**

Extreme ranges of acidity or alkalinity, oil and grease, cadmium, nickel, zinc, copper, lead, selenium, iron, chromium, mercury, arsenic, cynides, fluorides, phenol, silicate minerals, pesticide residues, vinyl chloride, formaldehyde, CO, SO_2, NH_3, H_2S, chloro-fluoro carbons, HCFC, methyl bromide, radio active minerals etc. are major harmful chemicals.

Each and every chemical is likely to result in chronic or acute effects depending on its threshold limit.

5. **Is there a relationship between social development and community health?**

Provision of schools, hospitals, communications, power supply and other facilities definitely improve the living conditions in a community as poverty (malnutrition, inadequate housing, lack of individual latrines, protected water supply and sanitation measures) exposes people to communicable diseases. However all the welfare measures do not reach some sections of the population for a number of reasons. When some people are unhappy, unity in the total community system collapses.

Illegal trade activities are responsible for deforestation, reduction in the areas of water bodies and indiscriminate disposal of wastes. Adulteration has increased in the consumer market as there is a change in the moral values in the society.

Globalization increased the opportunities on one hand but the individuality is lost in wilderness.

REVIEW QUESTIONS

1. Explain the terms population density, immunization and sanitation.

2. Write a note on the various welfare programmes taken in a developing society.

3. Justify the need for controlling population growth rate?

4. Establish the relationship between environmental pollution and human health.

5. Outline the sanitation measures to be taken during a kumbh mela.

6. In what way, information on AIDS, is useful to a student?

OBJECTIVE TYPE QUESTIONS

Fill in the blanks with suitable words:

1. In the last decade, agricultural land had/had been ____________ (ha/head).

(a) decreased (b) increased (c) constant (d) fluctuating

2. Better standards of living are expected with

(a) literacy (b) health care (c) women welfare (d) all of the three

3. Preferred population growth pattern is

(a) sigmoid (b) exponential (c) arithmetic (d) none of the three

4. Through blood, ____________ disease is transmitted.

(a) AIDS (b) HIV (c) Malaria (d) Cholera

5. Filaria is a/an ___________ disease.

(a) epidemic (b) endemic (c) sporadic (d) common

Part Two

POLLUTION CONTROL AND ENVIRONMENTAL MANAGEMENT

CHAPTER 6

POLLUTION AND CONTROL

TOPICS COVERED

- Air Pollution
- Water Pollution
- Solid Wastes Disposal
- Soil Pollution
- Marine Pollution
- Thermal Pollution
- Noise Pollution

6.1 AIR POLLUTION

Clean air contains 78.09 per cent nitrogen, 20.94 per cent oxygen and 0.31 per cent carbon dioxide by volume. Argon, neon, helium and methane are also present in very small quantities. Human activities release a number of toxic chemicals, particulate matter and gaseous emissions into the ambient atmosphere. These substances disperse due to wind movements in horizontal and vertical directions. They may accumulate, react or dilute depending upon the prevailing conditions. These impurities are grouped as follows:

(i) Particulate matter (dust, smoke, fumes, droplets)
(ii) Inorganic substances (CO, NO_x, SO_x, H_2S, HF, NH_3, Cl_2)
(iii) Organics (methane, benzene, formaldehyde, chlorinated hydrocarbons)

Air pollution means presence of one or more contaminants, in excessive concentrations (than maximum accepted and permissible limits), for sufficient duration, causing harm to human beings, plants and animal life, damage to the materials and interference to the normal activities.

Sources and Characteristics

Industries, power plants, foundry units, etc. are stationary sources, limiting the emissions to a particular area only. Vehicular transport systems on roads and railways are mobile sources of pollution. They are line sources mostly affecting the trees and houses along the movement path.

Natural Sources

Volcanic eruptions, forest fires, storms, soil erosion, electrical discharges, oxidation, condensation, polymerization, photochemical and other natural reactions in the atmosphere contribute contaminants into the ambient air. Oxidation of methane from decaying vegetation results in the formation of carbon monoxide. Use of fertilizers and pesticides in agricultural operations leave residual chemicals like arsenic, lead, fluorides etc. in the top soil. Salt sprays from ocean, spores and pollen from plants and weeds, smoke from fuel burning, CO_2 from respiration, H_2S and CH_4 from anaerobic decomposition of wastes are common natural sources of air pollutants. Radioactive minerals from the earth's crust are also very harmful pollutants.

Anthropogenic Sources

Human activities such as mining, transportation, combustion, industrial processes, agricultural activities, release gaseous and particulate matter into the atmosphere.

(i) Combustion produces CO, NO_x, smoke, particulate matter, hydrocarbons, lead and other impurities, depending upon the quality of raw material, process of combustion, ignition temperature, air-fuel ratio and efficiency of burning. Fuels containing sulphur as an impurity releases SO_2 on combustion. Some metallic oxides may also be present. Burning of solid wastes also contributes to smoke and nuisance value.

(ii) Various processing industries release pollutants based upon the manufacturing process and usage of raw materials (Table 6.1).

Table 6.1 Pollutants released by the industries

Industry	*Pollutants*
Cement	Cement dust, oxides
Paper	Mercaptons, H_2S, SO_2
Thermal power plant	Fly ash, SO_2, NO_x
Iron and steel	Dust
Metallurgical units	Dust, fumes, smoke
Petrochemicals and refineries	HC, CO, dust
Paints	Nitrobenzene, thinners
Asbestos	Fibres

(iii) Movement of vehicles release oxides of carbon, nitrogen and hydrocarbons from their exhaust emissions, in addition to smoke and particulate matter. Lead is a toxic substance in petrol driven automobiles.

(iv) Earthmoving machines, construction work and a number of engineering activities also cause air pollution. Collision, abrasion and grinding operations release dust and particulate matter.

Classification of Pollutants

Primary pollutants are those which can be emitted directly from the source. CO, SO_2, NO_2, HC, H_2S, HF, NH_3, particulate matter are all primary pollutants released from the source of generation.

Major primary pollutants in air are: CO (45%), SO_x (13%), NO_x (10%), HC (12%), Particulates (15%) and others (5%). (Figure 6.1)

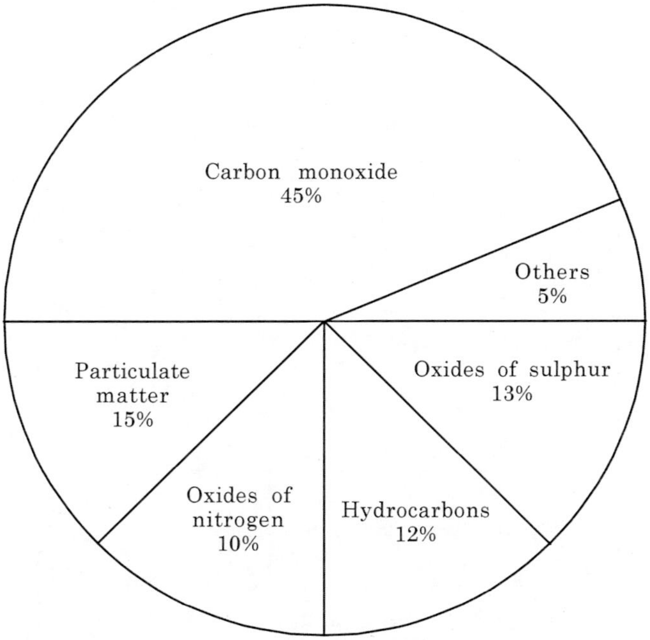

FIGURE 6.1 Primary pollutants in the atmosphere.

Transportation is a major source of pollution and carbon monoxide is the main pollutant released into the atmosphere.

Secondary pollutants are formed in the atmosphere when some primary air pollutants react with one another or with other chemicals in the air. SO_3, HNO_3, H_2SO_4, O_3, PAN (peroxy acetyl nitrate) and aldehydes are secondary pollutants. HC and NO_2 undergo photochemical reactions in the presence of solar radiation to form ozone, PAN and aldehydes. Photo chemical smog at ground level is injurious to human beings.

Characteristics of Pollutants

The nature and mechanism of reactions involved with the major pollutants and their harmful effects are discussed as follows:

Particulates or Dust

Solid particles generated by handling, crushing, grinding, spraying and other operations are known as *particulate matter*. Particles of different sizes from 2.5 microns to 100 microns appear in the atmosphere. Small and light particles will be in suspension, whereas bigger sized particles settle by their own weight. Fine particles reach human lungs and cause respiratory diseases. Size ranges of particles from various sources are shown in Figure 6.2. Fly ash is an ash content entrained in the combustion products after burning coal.

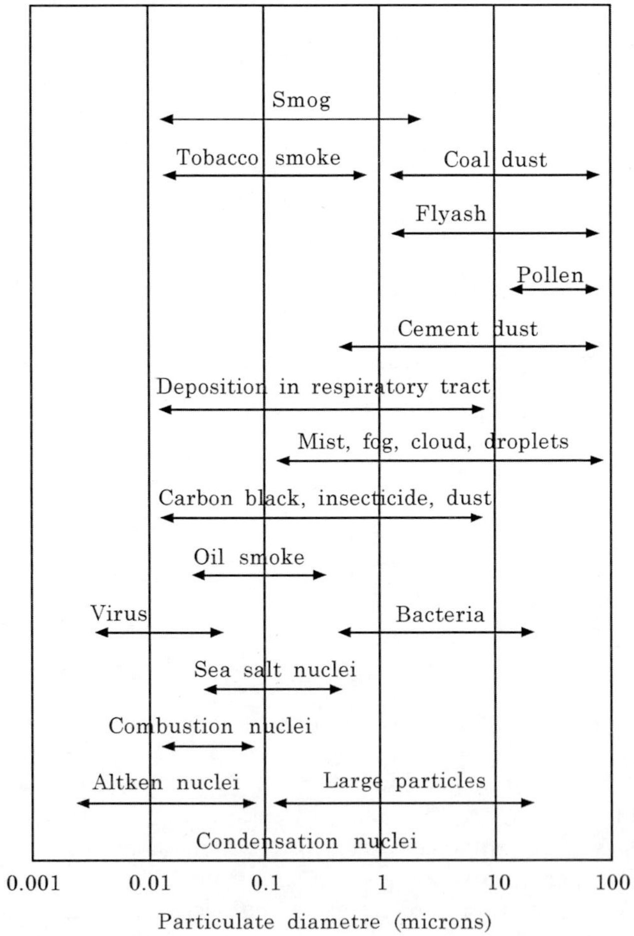

Sizes of atmospheric particulate matter

FIGURE 6.2 Particulate matter-size distribution.

Suspended particulate matter is generated from burning coal in power and industrial sector, burning petrol and diesel in vehicles, construction and agricultural activities. Arsenic, lead, nickel are toxic chemicals present. All air borne colloidal suspensions are aerosols, behaving as gases and remain for a long time. Mists are suspended liquid droplets. Fumes are solid particles generated by condensation from gaseous state. Fog is an aerosol of liquid droplets near the ground usually occurring during sunrise and sunset. Smoke is unburnt carbon particles. Smog is a mixture of smoke and fog. Smog reduces the visibility and is also responsible for accumulation of impurities in the atmosphere. In the presence of sunlight, photochemical reactions take place resulting in the accumulation of hazardous compounds at ground level. Particulates are classified as inorganic dusts, organic matter, droplets, smoke, fly ash depending upon their characteristics. Different particles behave in different ways. Their particle movement, rate of cooling or evaporation, scattering of light, settling rate, interaction and resistance, adhesion, sorption, migration, diffusion and other properties depend upon the particle size and nature. Particle distribution, mass concentration, chemical composition and transport behaviour help in deciding the process of removal of pollutants from air.

Gaseous Pollutants

Most common gaseous impurities are oxides of carbon, oxides of nitrogen, oxides of sulphur and hydrocarbons.

(i) Carbon monoxide is a colourless, odourless gas formed during incomplete combustion of carbon containing fuels. Coal mining and oxidation of methane also produces CO. Formation of CO depends upon factors like air-fuel ratio and ignition temperature. Autoexhaust contains CO and CO_2. Soil, vegetation, sea water and microorganism remove CO from the atmosphere.

(ii) Nitrogen dioxide (NO_2) is a reddish brown irritating gas, released from fuel combustion in industrial plants and automobiles.

$$N_2 + O_2 \longrightarrow 2NO$$
$$2NO + O_2 \longrightarrow 2NO_2$$

NO_2 emissions can be reduced by reducing peak combustion temperatures, residence times and air availability during combustion. Toxic secondary pollutants are formed in the presence of NO_x, HC and O_3.

(iii) Sulphur dioxide is a colourless and pungent gas obtained from the coal burning process, containing 10–15 per cent of sulphur as an impurity in coal. Metals are also extracted from sulphide ores and SO_2 is released from the processing units. SO_2 combines with water to form sulphuric acid which is also harmful to life and materials (see Table 6.2).

Table 6.2 Air pollutants and its effect

Pollutants	*Effects on human health*
Dust	Cough, cold, asthma, bronchitis, allergy, T.B.
SO_2	Respiratory irritant, effects lung function.
NO_x	Inflammation of lungs.
CO	Reduces oxygen carrying capacity of blood.
H_2S	Causes nausea, throat and eye irritation.
CO_2	Asphyxiation.
HC	Highly reactive.

In addition, Cd, Hg, Mn, Cr, Vd, F and As are toxic chemicals which are injurious to plants, human beings and animals. Plants are more sensitive when compared to animals and human beings.

Vegetation

Dusts and aerosols are deposited on leaves, affecting the rate of photosynthesis and transpiration. This results in premature dropping of leaves, reduction of chlorophyll, damage to leaf tissue, ultimately leading to plant decay. Fluorine, ozone, SO_2 and arsenic are also injurious to plant life.

Materials

Air pollution causes significant loss to the materials and property due to corrosion, abrasion, deposition and chemical reactions.

Photochemical Smog

Reactive hydrocarbons interact with ozone to form a hydrocarbon free radical, which readily reacts with oxygen to form another free radical (RCH_2O_2). This reacts with NO to produce NO_2 and free radical RCH_2O. When RCH_2O reacts with oxygen, a stable aldehyde RCHO and hydro peroxyl radical HO_2 forms which is extremely reactive and regenerates the hydrocarbon free radical. This goes as a chain reaction. Finally Peroxy-Acetyl Nitrate (PAN) which is a potent and toxic irritant is formed due to the presence of particulates, NO_2, O_3, HC (volatile organic carbon) and CO in the atmosphere (sunlight).

Others

Ozone (O_3) is a highly reactive gas with unpleasant odour. H_2S (a foul smelling gas) comes from anaerobic degradation of waste matter. HF, HCl and Cl_2 are acidic in nature. Large quantities of hydrocarbons also enter the atmosphere which have high residence time. CCl_4, CH_3Cl, $CHCl_3$, and C_6H_6 are more hazardous air pollutants.

Harmful Effects of Air Pollution

Air pollution affects health of the human beings and animals, agricultural yield, materials, property, movements, tourism and reduction in visibility in the atmosphere. Extent of damage depends upon the nature of the pollutant, concentration and duration of exposure. Atmospheric conditions like temperature and humidity also play a prominent role (see Table 6.3).

Table 6.3 Effects of air pollutants

Material	*Pollutants*	*Effects*
Stone, marble, concrete	SO_2, H_2SO_4, HNO_3, dust	Surface erosion, discolouration.
Metals	SO_2, H_2SO_4, HNO_3, dust	Corrosion, tarnishing, loss of strength.
Ceramics	HF, dust	Surface erosion.
Paints	SO_2, H_2S, O_3, dust	Surface erosion, discolouration and soiling.
Paper	SO_2	Embrittlement, discolouration.
Leather	SO_2	Surface deterioration, loss of strength.
Textiles	SO_2, NO_2	Fading, soiling and deterioration.
Rubber	O_3	Cracking and loss of strength.

Air pollution causes damage mainly to the respiratory tract. Irritation of nose, eyes, throat and skin; cough, nausea, suffocation, breathlessness and allergy are common symptoms. Some chemicals also enter the nervous system while some damage heart, kidneys and liver. Bronchitis, asthma, dermatitis, T.B., lung cancer may also occur due to air pollution. Disasters may happen when exposed to very high concentrations even for a short interval of time. Continuous exposure even to low concentrations of silica or asbestos dust in the case of workers cause respiratory diseases like silicosis or asbestosis. All chemicals have a threshold limit value beyond which the chemical will be toxic to human life.

Environmental factors like temperature, sunlight and humidity also play a crucial role in reducing the strength, quality, utility and value of any material.

Visibility

Pollutants in the air absorb light. Visibility gets reduced when fog traps aerosols and particulate matter in the air especially during sunrise.

Air Quality Standards

To identify whether a particular area is polluted or not, air sampling has to be done, to measure the concentration of various pollutants and then compare with the maximum permissible limits prescribed by Bureau of Indian Standards and Pollution Control Boards based upon extensive epidemiological surveys (see Table 6.4). Proper care should be taken with regard to number of sampling stations, duration of sampling, methods of sampling and analysis to determine the concentration of pollutants.

Table 6.4 Different pollutants and permissible levels

Pollutant	*Concentration in ambient air in $\mu g/m^3$*			
	Duration	*Sensitive areas (hospitals, schools)*	*Industrial areas*	*Residential areas*
SO_2	Annual	15	80	60
	24 hours	30	120	80
NO_2	Annual	15	80	—
	24 hours	30	120	
Suspended particulate matter	Annual	70	360	140
	24 hours	100	500	200
Lead	Annual	0.5	1.0	0.75
CO	24 hours	0.75	1.5	1.0
	8 hours	1.00	5.0	2.0
	1 hour	2.00	10.0	4.0

Pollutant levels in three different cities are presented as follows:

Pollutant	*Concentration of pollutant*					
	Delhi		*Mumbai*		*Calcutta*	
	tonnes/year	$\mu g/m^3$	tonnes/year	$\mu g/m^3$	tonnes/year	$\mu g/m^3$
SO_2	46,000	(253)	1,57,000	(33.7)	25,500	(48.6)
S.P.M.	1,16,000	(446)	50,000	(3.31)	2,00,000	(37.6)
CO	2,65,000	—	1,88,000	—	1,77,000	—
NO_x	73,000	(43.5 mg/mL)	58,000	(35)	40,200	(34.1)

Pollution Sampling and Analysis

The main purpose of sampling is to know the air quality or the nature and degree of pollution so that health hazards can be prevented. Air samples may be taken from the ambient atmosphere, automobile exhaust or from an industrial chimney.

(i) Settleable dust quantity is measured by collecting dust in a jar of 100 sq. cm. opening, kept on the ground and open to the atmosphere for 30 days. The mass of dust collected may be converted into tonnes per sq. km.

(ii) Suspended particulate matter is taken into high volume air sampler (Figure 6.3) through a glass fibre filter paper. Suction motor works at constant rate (litres per minute) for a fixed time. Concentration of suspended particulate matter is calculated on the basis of collection on the filter paper and is expressed as mg/m^3 of air for fixed time period.

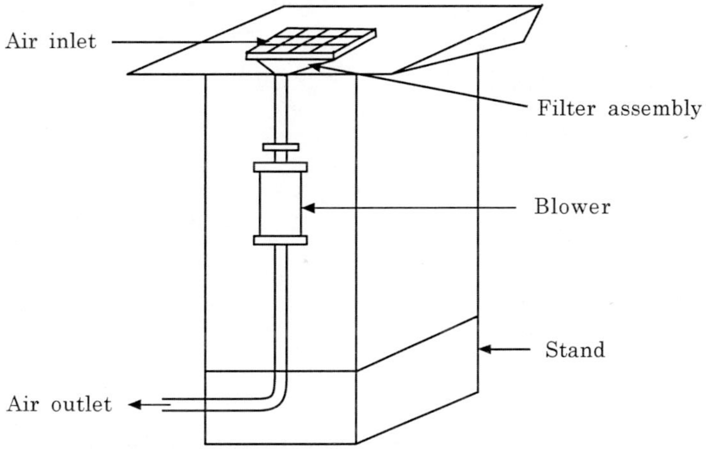

FIGURE 6.3 High volume air sampler.

(iii) Gaseous pollutants are collected through a sampling train (Figure 6.4) using different absorbing solutions for different pollutants which may be analysed in the laboratory. Air sampling from a stack requires proper probe as the air flow is at a high velocity, temperature and pressure.

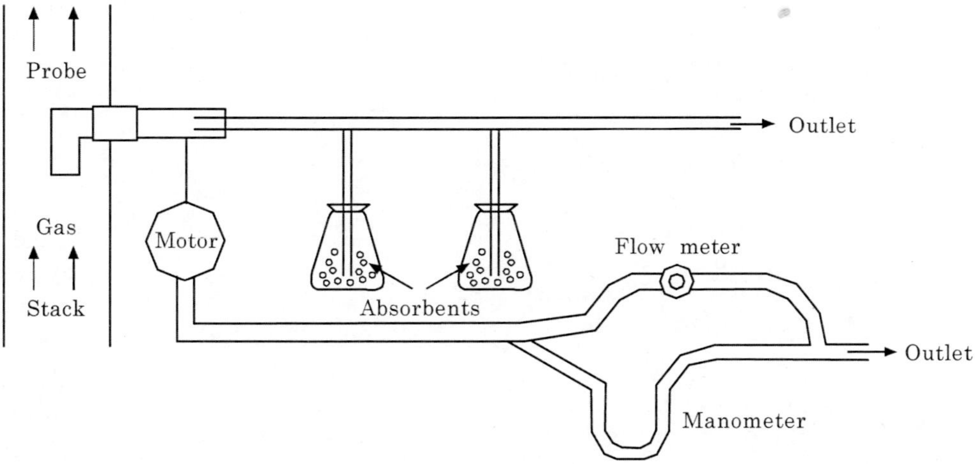

FIGURE 6.4 Sampling train.

Meteorology

Atmosphere above the surface of earth consists of several layers. The average temperature and pressure of the atmosphere varies with altitude. Under normal atmospheric conditions, temperature decreases with height (lapse rate) upto an altitude of about 15 km. Warm air moves up. If warm air is trapped under a cooler layer (i.e. inversion condition), pollutants will not get dispersed away from the source of generation. The direction and velocity of wind plays a crucial role in the dispersion of pollutants. The average wind direction and velocity of wind may be depicted in the form of a wind rose (Figure 6.5).

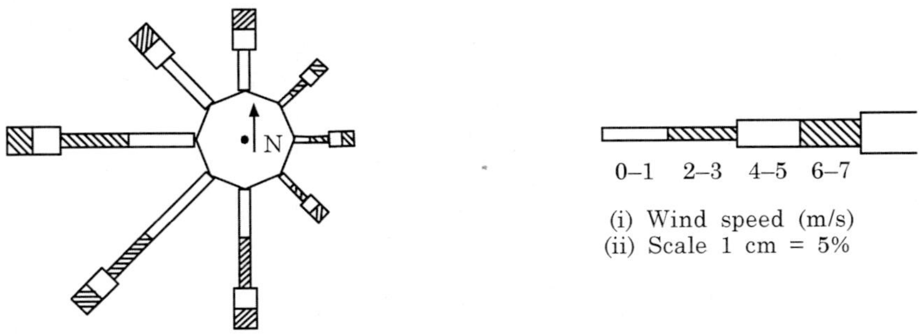

FIGURE 6.5 Wind rose diagram.

Dispersion is the process of spreading of pollutants in different directions in the atmosphere under different meteorological conditions of stability. Pollutants released from industrial stacks, gets mixed up in the air between 100 to 500 metre height above ground level, depending upon the meteorological factors prevailing in the local area. Concentration of pollutants decrease with the lapse of time and with distance from the source, in the direction favoured by the wind movement.

Removal of Particulate Matter

Air pollution may be minimized by the following ways:

(i) Selecting proper raw materials
(ii) Modifying the equipment
(iii) Changing the process reactions
(iv) Regulating or recovering the wastes
(v) Changing the operational schedules
(vi) Using proper control devices

Smoke and sulphur dioxide can be reduced by selecting low volatile low sulphur coals. Double catalysis double absorption process reduces SO_2 emissions in the production of sulphuric acid. Use of proper combustion devices, maintaining proper air fuel ratio, ignition temperature, stack design and maintenance, reduce the output of pollutants. All these attempts are successful to a reasonable extent. Particles vary in size, shape, distribution and characteristics. The choice of equipment depends upon a number of factors such as:

(a) Physical and chemical properties of the particles
(b) Particulate size, concentration and volume of impurities present
(c) Temperature and humidity of the medium
(d) Carrier gas volume and rate of removal
(e) Efficiency required
(f) Economic considerations

Settling chambers, cyclone separators, wet scrubbers, filters and electrostatic precipitators are widely used in the industry for the removal of particulate matter.

Air flows at a low velocity, giving sufficient opportunity for settling of larger size (> 40 m) particles due to gravity forces only in a settling chamber (Figure 6.6).

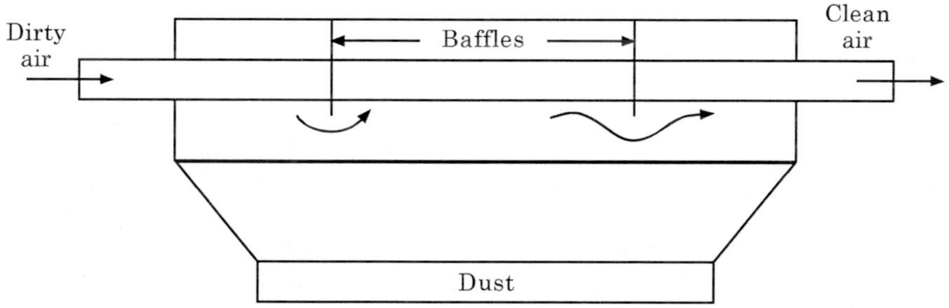

FIGURE 6.6 Settling chamber.

A particle settles depending upon the horizontal flow velocity, upward forces, resistance due to buoyancy and downward gravitational force on the particle. The more the particle size, the more is the settling rate. Settling chambers are designed based upon the minimum size of particles to be settled and choosing such a flow rate and surface area loading, to achieve the objectives effectively. Dust removal may be continuous or at different intervals of time. A sudden reduction in velocity helps in a better removal condition in a baffle chamber. The advantages are: low initial and maintenance costs and dry collection of particles.

In a cyclone (Figure 6.7), gas movement generates a centrifugal force and throws the particles towards the wall. Clean gas moves up as particles slide downwards for collection. Efficiency increases when the diameter of the cyclone is less. Performance of a cyclone is affected by:

(i) Temperature of influent gas
(ii) Variations in flow velocity and volume
(iii) Humidity conditions
(iv) Proportioning the (dimensions) parts
(v) Shape, size, specific gravity and influent particle concentration

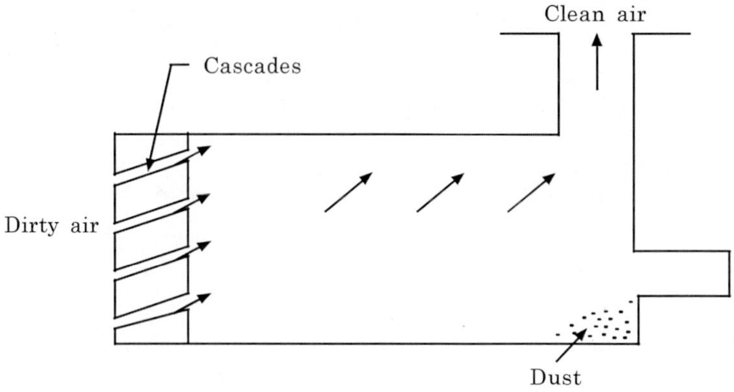

FIGURE 6.7 Cyclone concentrator (helical entry).

Bouncing and re-entrainment of particles, particle interactions, gas viscosity and density, may decrease the efficiency of a cyclone. Inside wall smoothness, formation of vortex, dust density and particle concentration help the collection efficiency. Wall surface may be affected by corrosion from reactive gases, sticky cake formation, or erosion due to attack from particles.

Cyclone separators may be used at high temperatures or pressures for wide range of particle sizes and concentrations. Multiple cyclones are large single compact units of smaller cyclones under parallel operation (simultaneously) to obtain higher efficiency of collection. Cyclones have an efficiency range from 60 to 90 per cent but may not effectively remove particles of less than 20 micron size. There are no moving parts and can be easily installed for handling large volumes of gas.

Wet scrubbers (Figure 6.8) are collecting devices where water captures dust particles. 0.1 to 20 micron size particles are removed in these units. Dry air movement and water sprays may be in the same, opposite (more optimum) or any direction [Figure 6.8(a)]. Wet collection devices are preferred in foundries, mining operations and metallurgical industries. In a simple spray chamber [Figure 6.8(b)] water flows in the downward direction and traps the impurities. The chamber may be packed for improving the contact opportunity and collection efficiency. A baffled chamber [Figure 6.8(c)] may also be used for wet impingement. In a cyclone scrubber [Figure 6.8(d)] polluted air is tangentially swirled around whereas water sprinkles into the chamber continuously. In a venturi scrubber [Figure 6.8(e)], high velocities (60–100 m/s) at throat section atomize the liquid. Particles strike against moving droplets because of the velocity difference between the droplets and particles, droplets accelerate in throat section. The action is not complete in throat section but continues even afterwards till settling. Hence throat length is important for travel and collection of particles. This type of scrubbing device is very efficient for the removal of particulates < 0.5 microns, due to a better capture between high velocity collision particles and atomised water droplets.

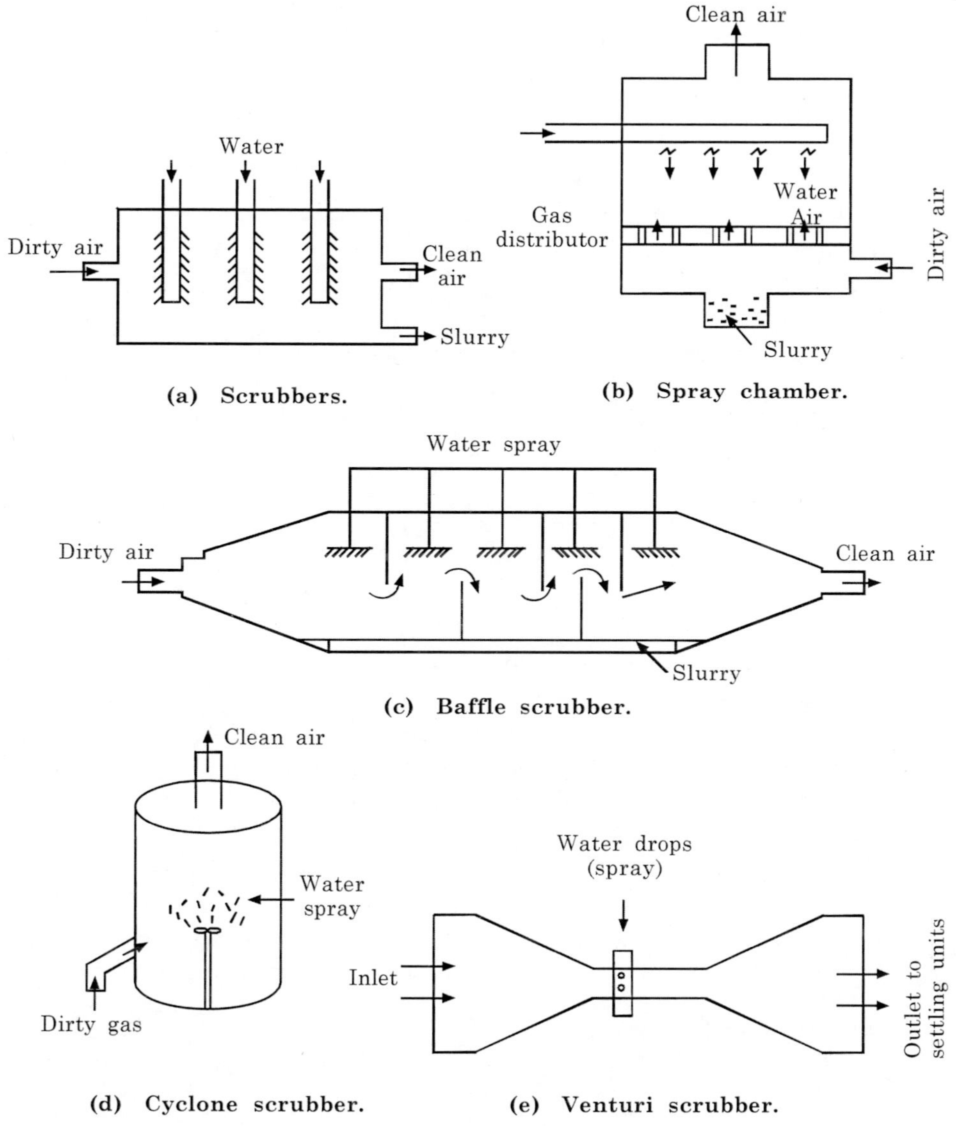

(a) Scrubbers. **(b) Spray chamber.**

(c) Baffle scrubber.

(d) Cyclone scrubber. **(e) Venturi scrubber.**

FIGURE 6.8 Wet scrubbers.

Advantages of wet devices are as follows:

(i) Reliable with high collection efficiency
(ii) Highly soluble gases like NH_3, HCl, HF are also removed
(iii) Removes corrosive chemical dusts that may be difficult for removal in filters or electrostatic precipitators

Bag filters, consisting of porous, fibrous material, retain the particulate matter when the dirty gas passes through the interstitial spaces of filter. Filter bags (Figure 6.9) are usually 15 cm diametre and 3 m long fibre glass material filters. Air enters from the bottom and flows out along its sides, leaving the dust to form a cake inside the filter.

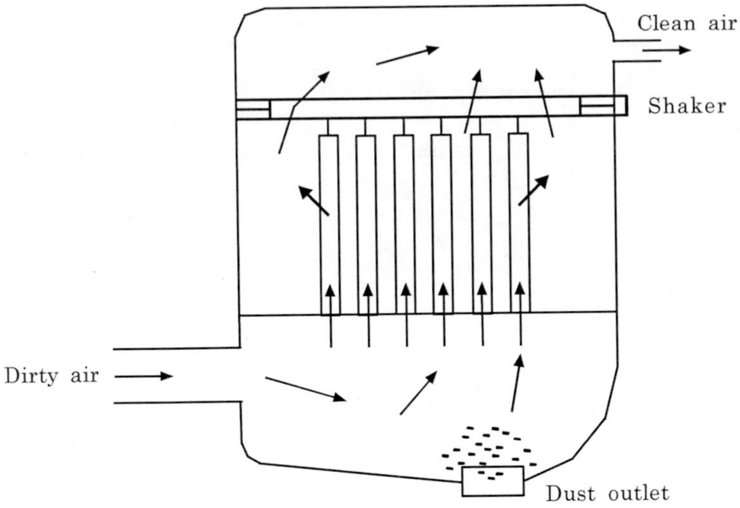

FIGURE 6.9 Bag filter.

Semi permeable wooven or felted material is used. The filter material depends upon the gas temperature and the nature of (acidity or alkalinity) gas as it has to resist abrasion and chemical actions. Common materials used are cotton (natural, fibre, cellulose), wool, nylon, orlon, teflon (fluorocarbon), fibre glass, and stainless steel. They remove very fine particles of 1 micron size with 95 per cent efficiency.

(i) Fine particles penetrate through fabric and also causes rupture of cloth which may be difficult for location, repair or replacement.
(ii) Humidity should be controlled as hygroscopic dusts create problems.

Electrostatic precipitators (Figure 6.10) are widely used in cement,

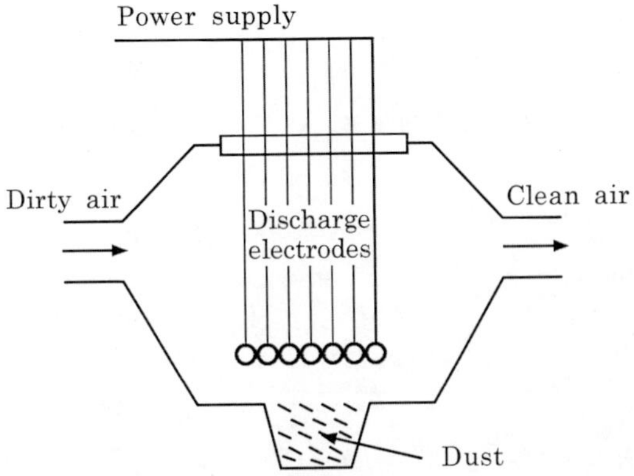

FIGURE 6.10 Electrostatic precipitators.

power and metallurgical industries. Effluent gas passes through a high electric field and gets ionized. Negatively charged particulates move towards positively charged collecting electrodes and attach to their surface. These precipitators are very efficient in removing all particulate matter.

The steps involved are as follows:

(i) Impart electrostatic charge to particles.
(ii) Create electric field in the flow region.
(iii) Particles develop force of attraction.
(iv) Particles migrate towards oppositely charged electrodes.
(v) Electrodes collect the impurities.
(vi) Particulate dust is removed by shaking or rapping electrodes.
(vii) Hopper below collects the dry particulate matter.

Basically two types are available—pipe type and plate type. In the pipe type, discharge electrode is a wire suspended (in a small pipe) with sufficient weights to maintain the position. Precipitation takes on the inner wall surface. In the case of plate type devices, collecting electrodes and discharging electrodes are placed in a parallel position. Corona forms between an active high voltage electrode such as a fine wire and a passive ground electrode such as a pipe or may be between two parallel or different plates. It is a mechanism of formation of ions in the gas. Dust particles carried through this space, instantaneously become highly charged and immediately migrate and attach to the collecting electrodes. Negative corona formation is more stable and efficient for collection.

Electrostatic precipitators have a number of mechanical parts and electrical components and so have a number of operational and maintenance problems. A number of operational problems appear with precipitators such as poor alignment of electrodes/plates, excessive dust in corona field, overloading, particle re-entrainment, insufficient high voltage field, highly resistive particles, presence of excess carbon monoxide in gas. The demerits of electrostatic precipitators are high capital cost and sensitivity to inflow variations.

Comparison of Various Devices

Choice depends upon the characteristics of impurities and the collection efficiency needed. Cyclones are (economical, easy to install and maintain) used when dust concentrations are fairly high. Wet scrubbers are used when fine particles are to be removed at high efficiency, and when moist (cool) conditions are not objectionable for collection. Combustible gases, particulate and gaseous impurities can also be removed. Fabric filters collect valuable material and are operated at low temperatures and volumes. Though efficiency is high, installation and maintenance costs are more. Electrostatic precipitators can handle large volumes of gas, at different temperatures and conditions. Efficiency is very high but requires constant attention and care.

Control of Gaseous Impurities

Absorption, adsorption, condensation and combustion are the main processes by which gaseous emissions are purified.

Selection of a suitable equipment depends on the considerations of the following factors:

(i) Volume of air and concentration of pollutant
(ii) Recovery of valuables
(iii) Simultaneous removal of particulates
(iv) Cost and efficiency considerations

Packed beds [Figure 6.11(a)] provide contact opportunity for effective absorption or adsorption [Figure 6.11(b)]. Water is a good solvent for absorbing

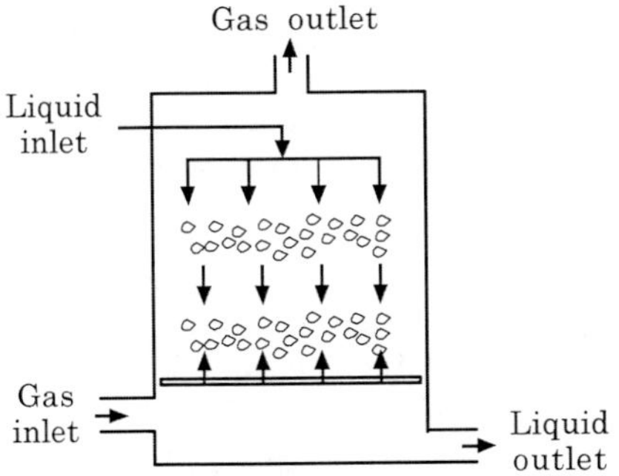

FIGURE 6.11(a) Packed bed absorbers.

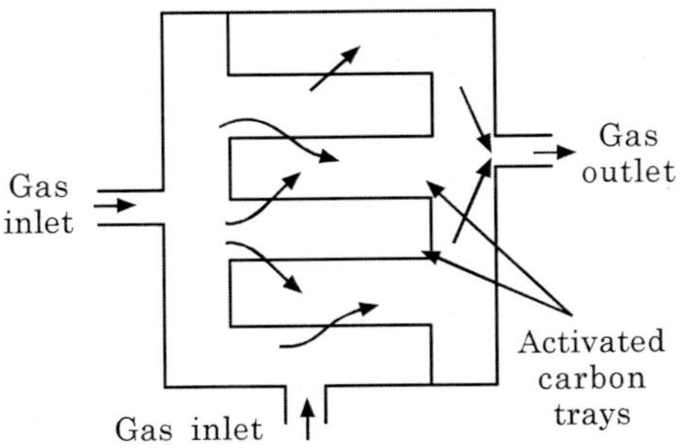

FIGURE 6.11(b) Packed bed adsorbers.

a number of gases. Adsorption is good for handling organic vapours. Activated carbon (manufactured using coconut shells, bituminous coal or petroleum residues), siliceous compounds (silica gels, fullers and diatomaceous earth, synthetic zeolites), alumina, anhydrous $CaSO_4$, Ca or Mg silicates are used as adsorbents. Finely powdered activated carbon removes gases and bad odours too. Lime stone or alkalised ammonia is used for the removal of SO_2. Iron oxide bed removes benzene and H_2S. Silica gel adsorbs oxides of Nitrogen and Bromine. Regeneration of chemical and material recovery is possible.

Condensation of a vapour from dirty air by circulating coolants in the tubes (Figure 6.12), helps in the removal of hydrocarbons.

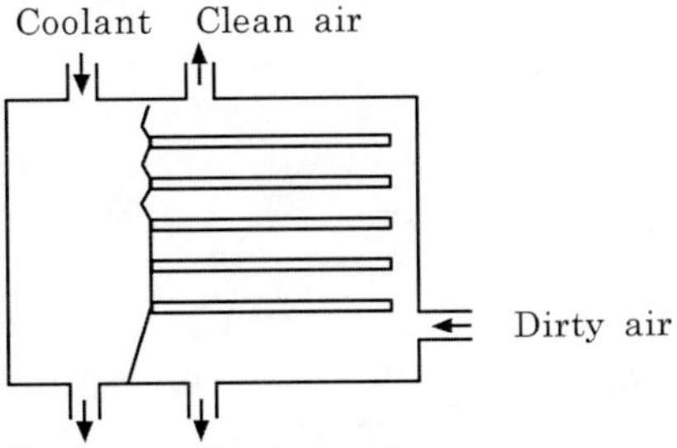

FIGURE 6.12 Condensers.

Incinerators (Figure 6.13) are simple, safe and reliable systems used for combustion of variety of gases. If required, fuel may be added or catalyst may be used to ensure complete combustion.

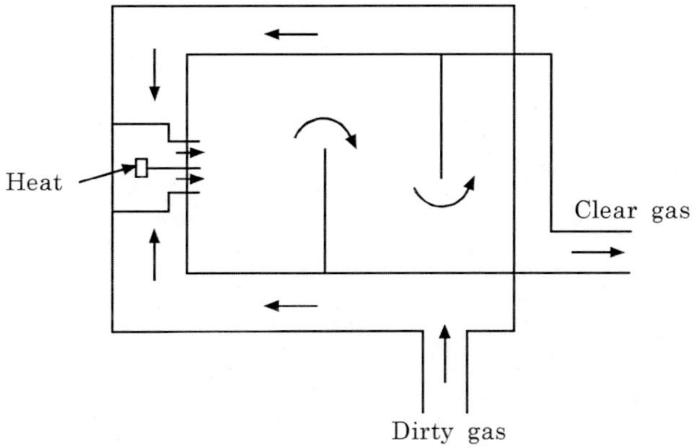

FIGURE 6.13 Thermal incinerators.

Air Pollution Control in Industries

(i) *Cement industry.* Dust is generated at all stages of the plant (production and distribution). Concentration of air borne dust from limestone crushing, kiln firing, clinker zone, coal mills, packing and transportation units is around 20,000 parts per cubic centimetre. SO_2, NO_x, CO_2 are the other major primary constituents of emissions.

Dust may be collected using cyclones or filter bags in crushing area or in grinding area of cement clinker units. Electrostatic precipitators are generally used to collect dust from stack emissions.

(ii) *Metallurgical units.* Coal and coke handlings, charging, processing units are responsible for emission of pollutants in iron and steel units. They are mainly hydrocarbons and coal dust. Particulates also come from sintering unit. Blast furnace or open hearth units, rolling, processing and foundry operations are also sources for creating potential emissions. They may be oxides of metal, coal ash, or scrap material. Toxic metallic compounds may be released from various non-ferrous metallurgical operations as particulates and fumes. Dusts are also associated with mining and smelting units.

(iii) *Petroleum products.* Crude oil is distilled, separated, converted and treated in various ways to get finer petroleum products. Smoke, particulate matter, hydrocarbons, CO, SO_2, NO, NH_3, are the main pollutants from the emissions, depending upon the refining process.

Vapours may be removed from dirty gas through condensers, scrubbers or adsorbers. Cyclones may be used for the removal of particulate emissions.

(iv) *Power plants.* Combustion of fuel causes significant amounts of air pollutants. Fly ash, SO_x, NO_x are the primary pollutants entering into the atmosphere. CO and unburnt HC are also emitted. Dust comes from coal units.

Electrostatic precipitators are used to remove various particulate emissions.

Indoor Air Pollution

Indoor air pollution results mostly when ventilation is not proper. Fuel combustion, cooking vapours, cleaning sprays, solvent evaporation from insecticides, cosmetics, paints and varnishes, aerosols from consumer products like deodorants and cigarette smoking, ozone from photo copying and laser printer, fungus and mites from beds, allergens from house dust, Formaldehyde from furniture stuffing, particle board from insulation, methyl chloride from paint thinners, para dichlorobenzene from air fresheners, nitrogen and carbon oxides from stoves and Radon (Ra) from radio active soil and rock surroundings cause irritation and respiratory diseases. Fine particles which contains toxins and metals such as lead and cadmium are higher than outdoor pollution level. As people stay and are exposed for more time to the pollutants, health risks are more. All these pollutants cause a sick building syndrome.

Vehicular Pollution

In urban areas, air pollution levels have increased due to the ever increasing number of vehicles on the road. CO, HC, Pb, NO_x are the emissions from petrol driven vehicles whereas smoke, NO_x, SO_2 and traces of CO and HC emanate from diesel vehicles. Benzene and polycyclic aromatic hydrocarbons are released from all types of vehicles. Statistics reveal that 90 per cent of pollution load is from petrol driven vehicles only and CO and HC are the main pollutants.

The following measures reduce the pollution to a good extent:

(i) Increase the width of the road and plant trees on the road margins.
(ii) Improve the efficiency of public transport system.
(iii) Ensure smooth traffic flow so that average speed of a vehicle is 40 to 50 km/h to ensure minimum emissions.
(iv) Adopt modern technology to modify carburettor, ignition devices or catalytic converter to ensure complete combustion.
(v) Traps may be used to control smoke from exhaust systems.
(vi) Use alternate fuel like compressed natural gas.
(vii) Ensure implementation of vehicular emission standards as per Motor Vehicles and Environmental Protection Acts.

6.2 WATER POLLUTION

Water Supplies

Rain water flows over the surface of the earth to form lakes and rivers and a good amount of water slowly percolates into the ground layers. Lakes, ponds and reservoirs are water bodies supplying drinking water. Rivers carrying water through towns and villages, serve as a source for drinking water. Water is also available below ground level due to storage in subsoil region. Rain water collects dusts and gases from the atmosphere. Surface water collects more of suspended, colloidal particles and more dissolved oxygen. Ground water dissolves more minerals and salts. Water quality depends upon the catchments area and characteristics of a river basin (Figure 6.14). Depending upon their nature, impurities present in water may be classified as follows:

(i) Physical, chemical and bacteriological
(ii) Suspended, colloidal and dissolved
(iii) Organic and inorganic
(iv) Mineral salts
(v) Toxic chemicals

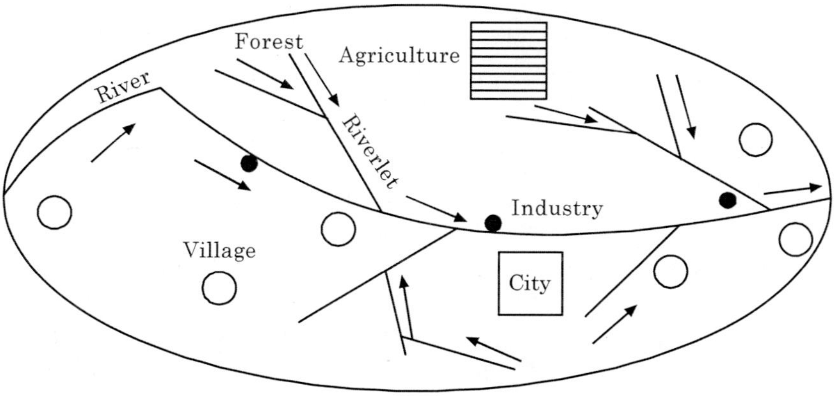

FIGURE 6.14 River basin.

Rubbish from town, pesticides and fertilizers from agricultural fields, domestic and industrial waste waters may join any water course. That is, water from any source is likely to have impurities in some form or other and there is no pure water anywhere. But all substances present in water do not necessarily cause harmful effects. Drinking water must satisfy the following requirements:

(i) Safety from water borne or associated diseases should be ensured.
(ii) Wholesomeness which means that needed salts and chemicals should be present at the required concentration levels.
(iii) Water should be clear, palatable, fresh and free from excessive concentration of undesirable chemicals.

(iv) Water must be free from all toxic chemicals that are injurious to various forms of life.

Water is used for various purposes such as:

(a) Domestic (drinking, cooking, cleaning, washing, bathing, flushing toilets etc.) use requires about 100 L/h/d (Litres/head/day)
(b) Commercial (hotels, cinema halls, bakeries, parlours) use requires about 45 L/h/d
(c) Public (schools, parks, hospitals, markets, railway stations, bus stands, etc.) use requires about 30 L/h/d
(d) Fire fighting needs about 2–5 L/h/d as a reserve against any contingency
(e) Losses and wastage may account for about 5–10 L/h/d.

The total demand (average) is about 200 L/h/d. This per capita demand may vary from 100–300 L/h/d, depending upon factors like climate, hours of water supply, standards of living, drainage systems, method and extent of charging the consumer and pressure available in the distribution system. Water consumption varies hourly, daily and seasonally.

Water Quality Parameters

(i) Physical characteristics: Colloidal and dissolved substances, organic and inorganic compounds, decaying vegetation, dyes, impart colour to the water. Minerals, phenols, chlorine, algae, decomposed organic matter may create typical, unpleasant and undesirable tastes to water. Surface run off may carry suspended solids (clay, silt, oils, grease), which often interfere with chemical reactions and microbial activity. Fine colloidal particles makes water turbid. They cause objectionable tastes and colours and also affects photosynthesis process. Palatability, solubility, viscosity and chemical quality of water is influenced by temperature.

(ii) Chemical characteristics: Acidity is caused by CO_2, mineral acids and salts from strong acids. Alkalinity is due to carbonates, bicarbonates and hydroxides. Aquatic life is more sensitive to pH. Chemical reactions and processes depend upon pH value.

Hardness is due to sulphates, chlorides, bicarbonates of calcium and magnesium in solution. Hard water prevents lather formation and causes scale formation. Fluorides, chlorides, sulphates, nitrates and toxic chemicals cause damage to internal organs of the human body.

Water contains a dissolved oxygen content of 8–10 mg/L. It is essential for aquatic life. Depletion of dissolved oxygen indicates utilisation by organic and inorganic compounds, i.e. contamination from domestic and industrial wastes.

(iii) Microorganisms: Pathogenic bacteria contaminate water sources and cause water borne diseases, like typhoid, cholera, gastro enteritis.

Based upon drinking water requirements, the permissible limits (Table 6.5) have been prescribed by various agencies like Bureau of Indian Standards and World Health Organisation. It should be noted that 2 or 3 litres of water is required per head per day for drinking purpose whereas the total requirement for all domestic uses put together is around 150 to 200 L/h/d. The total quantity of water is treated and supplied to each and every customer through a single pipe line only, as separate pipe lines are not economical and feasible.

Table 6.5 Drinking water quality standards

Parameters	*Desirable permissible limits* (mg/L)
Colour	5–10 (Hazen's scale)
Turbidity	10 (Nepheloturbidity units)
Taste and odour	Acceptable
pH	Neutral or slightly alkaline
Total dissolved salts	1000
Hardness as $CaCO_3$	200
Chlorides	250
Sulphates	150
Fluorides	0.8 to 1.5
Nitrates	45
Fe	0.1
Hg	0.001
Phenol	0.001
Cr, As, Cu, CN, Se, Cd, Pb	Each less than 0.01 to 0.1
Coliform count (MPN)	Nil
Radioactivity	Nil
Alkyl benzene sulphonate (detergents)	1

Water quality requirements for agricultural or specific industrial use is different.

Domestic Sewage

Sewage contains (99 per cent of water and 1 per cent solid by weight) mostly faecal matter and other solid wastes from a residential unit. More than 90 per cent of water supplied to a dwelling unit reaches a public sewer after use in domestic origin. Contaminants in waste water are listed below:

(i) Suspended solids and sludge deposits
(ii) Organic matter (measured by oxygen demand) which depletes dissolved oxygen from water sources
(iii) Dissolved inorganic solids
(iv) Pathogenic bacteria which are responsible for spreading communicable diseases

(v) Nutrients like nitrogen and phosphates that increase aquatic algal growth in water sources
(vi) Surfactants, phenols, heavy metal and toxic pollutants

Strength of sewage is sufficiently characterised by biochemical oxygen demand (BOD) and total solids.

Average composition of domestic wastes is likely to be as listed below:

Total solids	—	1,000 mg/L
BOD (5 days at 20°C)	—	200 to 250 mg/L
Total organic Carbon	—	200 mg/L
COD (chemical oxygen demand)	—	300 mg/L
Nitrogen	—	50 mg/L
Phosphorous	—	10 mg/L
Chlorides	—	150 mg/L
pH	—	7.5

Individual contribution is taken as follows:

BOD	—	80 g/day
Total solids	—	250 g/day
Quantity of waste water	—	150 to 250 L/day

Waste Water from Industries

Liquid wastes from various industries are generally disposed into the nearby environment with inadequate treatment. Pollutants generated by industries (Table 6.6) are as follows:

(i) Acids and alkalies
(ii) Materials that produce colour, turbidity and odour
(iii) Suspended and dissolved solids
(iv) Toxic and non-toxic inorganic chemicals
(v) Toxic and non-toxic organic chemicals
(vi) High temperature waste waters
(vii) Radio active materials
(viii) Bio-degradable and non bio-degradable substances
(ix) Oxygen consuming wastes

BOD of industrial wastes may vary from 1000 mg/L to 30,000 mg/L and pH value may be 3 or 10 (i.e. extreme conditions of acidity and alkalinity) whereas domestic sewage exerts only 200 mg/L of BOD and pH value at near neutral level.

Pollution can be reduced by the following measures:

(i) Disposal of cooling waters, floor washings, sullage waters is easy. Separation of liquid kiering wastes from textile mills and separation of chromium from metal finishing units are possible.
(ii) Number of useful substances like grease, caustic soda, silver, potash, chromium may be recovered easily for economical gains and to reduce pollution load.

Table 6.6 Characteristics of effluents from selected industries

Industries	*Characteristics*			
	BOD (mg/L)	pH	Solids (mg/L)	Others
1. Dairy	1,000	8	1,000	Biodegradable
2. Distilleries	30,000	4	50,000	Chlorides (5,000 mg/L), N (1,000 mg/L), Dissolved salts, Dark brown liquid
3. Fertilizers	500	8.5	2,000	N (10,000/mg/L), Chlorides (4,000 mg/L), Phosphates (1,000 mg/L), Sulphates (5,000 mg/L), F (10 mg/L) As, NH_3
4. Paper	400	8–9	1,500	Chlorides (300 mg/L), Fibrous matter, Lignin, Caustic soda, Brown colour
5. Pharmaceuticals	3,000 to 8,000	4 or 9	1,500 to 10,000	N (500 mg/L), Sulphides, Phosphates, Toxic organics, Strong odours
6. Tanneries	4,000	8–9	10,000	Chlorides (5,000 mg/L), High salt content, Sulphates, Chromium, Strong odour, Colour
7. Textiles	1,000	9–10	8,000	N (30 mg/L), Ca, Mg, Na, Zn, Sulphates, Colour

Note: Data is subjected to variation from plant to plant.

(iii) Waste water may be conserved by reusing or recycling within the industry, e.g. white liquors in paper industry, cooling waters in steel mills.

(iv) process and equipment modifications, depending upon the specific requirement, reduce the strength of waste. Use of phosphoric acid instead of sulphuric acid in pickling units, use of carboxymethyl cellulose instead of starch in textile mills, reduction of cyanide in metal plating industries etc., are quite useful steps in the production stages to reduce pollution.

Equalization and Neutralization

Treatment units are always designed for a specific influent characteristics and composition of wastes. Fluctuations or variations in the input quality affect the purification process. So uniform and consistent quality should be maintained by storing the liquid wastes for a certain period of time and mixing suitably the various components in a tank. From this unit, (i.e. an equalization tank or pond) uniform nature of the influent (with regards to BOD or solids) is fed into the treatment units.

Most of the biological treatment plants function very effectively at a range of 6–9 pH. So industrial wastes of acidic or alkaline nature are neutralized in a tank or pond. Sometimes waste waters from nearby units are mixed to obtain neutrality. Acid wastes may be passed through beds of limestone. Waste fuel gas as carbonic acid from boiler units, may be allowed to react with caustic wastes to obtain a pH value near seven.

When industrial wastes and domestic sewage are mixed for joint treatment, separate storage tanks help to properly proportion and regulate the quality and quantity of wastes into a purification plant. Industrial wastes may be partially treated and disposed into domestic sewers for further treatment.

Various Industries

Process operations, pollutant characteristics and purification procedures for various industries are presented briefly.

Dairy Plant

Mostly milk and their products, readily putrescible fats, proteins and lactose, strong organic and biodegradable substances come from dairy plants. High rate trickling filters and activated sludge plants are efficiently used, as aeration can reduce BOD to a good extent. Oxidation ponds are cost effective.

Cane Sugar and Distilleries Unit

Lime is added to crushed sugarcane. Juice is filtered, evaporated, and crystallised to obtain sugar. Mother liquor (molasses) is acidified with sulphuric acid, supplemented with nitrogen and phosphorus, fermented with yeast and finally distilled for obtaining ethanol. About 3,000 L/d of waste water is generated during the production of one tonne of sugar. Distillery wastes from molasses, containing high dissolved salts and ash content is acidic and has high BOD. Total waste water volume is 50 lit. per litre of alcohol produced, and originates mostly from stillage and fermenters.

Bagasse from sugar plant is useful as a fuel, and for preparation of fibre boards. Distillery waste is treated by lagoons followed by aerobic oxidation. Anaerobic digesters are useful as biogas and fertilizer (or cattle feed) can be obtained. Potash can be recovered from spent wash.

Fertilizer Industry

Fertilizer units are based on nitrogenous or phosphate compounds or a combination. Depending upon that, phosphates and nitrogenous compounds will come into waste waters. Fluorides, oil and grease, arsenic etc., also join the wastes.

500 litres of liquid effluent is approximately generated in the production of one tonne of urea fertilizer. In nitrogenous and phosphatic fertilizer industries, sulphuric, nitric and phosphoric acids, CO_2, ammonia, rock phosphates, coal, naphtha, etc., are used as raw materials. Carbon is removed for use again. Oil is removed by oil separators followed by coke absorption. Lime addition and clariflocculation neutralises and precipitates impurities like fluorides, and phosphates. Air stripping helps in the recovery of acids.

Ammonia and ethanol are stripped off in a distillation tower by means of steam. Air stripping at high pH, dissociates ammonia, and further converts ammonia into nitrates. It is anaerobically converted into nitrogen and finally sludge is removed in lagoons or settling tanks.

Paper Industry

Cellulose containing wood material is cooked in caustic soda, sodium sulphide mixture, under controlled temperature and pressure. Spent liquor (black liquor) is used for recovery of caustic soda, pulp is washed and bleached with chlorine. In the process, sulphate is reduced to sulphide and is converted to sodium hydroxide with lime treatment. That white liquor is sent to digestors. Pulp is used in paper manufacturing by adding dyes, alum and talc to fibres. Large volume of water is required to dilute pulp. Recovery of chemicals and fibres reduce the pollution load. Lignin is recovered from black liquor, after precipitating with lime addition and alum coagulation treatment. Sedimentation and flotation is a reliable treatment. Lagooning is an economical treatment for the removal of 80 per cent BOD. Activated sludge treatment is also a satisfactory treatment process for paper and pulp mill wastes. Nutrients (N, P) have to be added (if required).

Refineries and Petrochemicals

Processing and cooling operations generate a waste volume of 20 m^3 per 1 tonne of crude, in oil refinery unit. Major pollutants are oil [free, emulsified and dissolved (500 mg/L)] and chemicals such as acids, alkalies, sulphides (50 mg/L), phenols (50 mg/L), BOD (500 mg/L).

Depending upon the chemical process and manufactured item pH, BOD, solids, sulphides, phenols etc., are to be estimated. API gravity separators, sedimentation cum air flotation tanks remove oil substances. Coagulation is used for emulsified oils. Sulphides are removed by aeration. Biological (aerobic) treatment is good. Oxidation ponds remove BOD, sulphides, phenols and oils to a satisfactory level.

Air flotation and clarifloculators separate free as well as emulsified oil. Neutralization, activated sludge units and oxidation ponds are suggested units of treatment. Cyanides, fluorides and sulphides are segregated and treated. Sulphides are precipitated by the addition of ferric chloride. Fluorides are precipitated by the addition of lime or absorbed using activated alumina or activated alum treated carbon. Cyanides are removed by alkaline chlorination and later treated in aerobic biological units.

Pharmaceuticals

Basic drugs and various formulations are manufactured in pharmaceutical units. Waste water quantity and characteristics depends on the nature of products, processes, raw materials and purification process of finished products. Antibiotic plant yields a waste water with pH of 4.5 whereas sulpha drug unit results in waste water with pH of 9, BOD of 1,000 to 10,000 mg/L, and total solids exceeding 10,000 to 50,000 mg/L. Mostly, wastes come from filtration and distillation units. Neutralization, extended aeration, chemical oxidation, filtration, oxidation ponds are employed for treating pharmaceutical wastes. Chemical treatment with ferrous sulphate or alum is useful. Anaerobic lagoons are widely used. However, aeration units followed by clarifiers are also adopted in some cases. When total dissolved solids content is high, multiple effect evaporators are used. Hazardous wastes are incinerated.

Steel Plant

Wastes from coal washeries and blast furnace units are voluminous. A number of chemicals such as lead, zinc, chromates, phosphates, chlorides, sulphides, acids, alkalies, iron, silica, magnesium, phenols and cyanides may appear in the liquid effluent.

Froth flotation process removes fine coal particles. Clarifiers with or without coagulation units are used. Chemical oxidation and biological treatment are adopted.

Tanneries

Process operations include soaking raw hides in water treating with lime and sodium sulphide for removal of hair and flesh, treating hides in vegetable tannin solution of chromium sulphate liquor solution for tanning, washing, neutralizing and finally treating with emulsion of sulphonated oils and dyeing. Processes used are classified as pre-tanning, tanning and finishing operations.

Highly alkaline, high in suspended and dissolved impurities, with strong colour and foul smelling effluents come from different units. Segregation of chromium for recovery is desirable. Mixing of effluents from various units, screening, sedimentation, and conventional lagoons (aerobic or anaerobic or in combination), ASTP are practised. Chemical precipitation may cause sludge handling problems. Tricking filter treatment units and oxidation ditches are also very effective for reducing BOD.

Textile Units

Mills may produce cotton, woollen or synthetic fibres. Desizing, scouring, bleaching, mercerising, dyeing and printing operations of cotton mills generate liquid wastes to a tune of 3,000 m^3/day in a medium sized plant. pH of a cotton textile waste is 8–10 whereas it is 3–5 from a synthetic fibre unit. Organic, inorganic substances, starch, cellulose, sodium hydroxide, detergents, hypochlorites, dyes, pigments, sodium gums, waxes, sulphides, sulphates etc., are present in waste waters. Pre-treatment consists of segregation, recovery, recycling of water, screening, neutralization with lime, equalization to maintain flow balance and uniform characteristics, chemical coagulation with lime and iron salts and clarification for colour removal and finally biological treatment using trickling filter or activated sludge treatment plant appears to be the line of treatment.

Activities Concerned with Water Pollution

(i) A number of fertilizers and pesticides are used in all the agricultural operations. All these residual chemicals reach surface and ground water sources. Toxic compounds containing organics, mercury, chromium, calcium, nickel, lead and fluorides are very harmful to living organisms. Addition of nitrogen, phosphorous and other fertilizer chemicals cause eutrophication of lakes and water bodies. All these chemicals contaminate land leading to degradation.

(ii) When solid wastes are disposed by landfill operations, leachates reach water sources travelling through soil pores.

(iii) Nuclear and other hazardous wastes may join water bodies due to leakages, accidents, during storage, transportation, handling, treatment and disposal activities.

(iv) Misuse and abuse of water resources for bathing, washing, and other activities cause contamination and addition of detergents.

Water Pollution

Drinking water quality requirements and permissible limits for various physical and chemical parameters are given. If water contains impurities beyond the permissible level causing harmful effects on the users, then water is said to be contaminated or polluted. Sometimes, chemicals reach water source due to natural conditions such as rock minerals and salts getting dissolved in water sources as water is a good solvent (e.g. fluorides in ground water). This may be location specific only. But the main reason appears to be the indiscriminate disposal of wastes from domestic and industrial origin without adequate treatment.

(i) Tolerance limits for domestic waste waters to be discharged into inland surface waters are prescribed as BOD (5 days, 20°C) as 20 mg/L and suspended solids as 30 mg/L.

(ii) Tolerance limits for industrial effluents to be discharged into public sewers are prescribed as BOD (5 day, 20°C) as 500 mg/L, suspended solids as 600 mg/L, oil and grease as 100 mg/L and pH value to be 5.5 to 9.0.

(iii) Tolerance limits for industrial effluents (Table 6.7) to be discharged into internal surface water are given (Bureau of Indian Standards).

Liquid effluents after treatment can be used for irrigation purposes known as *sewage farming* (or land disposal) or may be disposed into large bodies of water such as rivers or sea. There is no third alternative for the disposal of waste water. Disposal on land or river causes water pollution unless effluents are properly treated. Wastes from various sources are not adequately purified before disposal. Land or river is not in a position to take the pollution load. Hence water pollution appears to be inevitable under these circumstances.

Table 6.7 Tolerance limits for industrial effluents

Parameter	*Tolerance limits* (mg/L except pH value)
BOD (5 day, 20°C)	30
Suspended solids	100
Oil and grease	10
pH	5.5 to 9.0
N	100
Fluorides	2.0
Phenolic compounds	1.0
As	0.2
Cd	2.00
Cr^{6+}	0.1
Pb	0.1
Hg	0.01
Zn	5.0
CN	0.2
Se	0.05
Ni	3.00

Harmful Effects

A number of chemicals if present in excess of permissible limits in drinking water affect human health in different ways. They are as follows:

Fluorides (softness of bones, fluorosis), Chlorides (sea water intrusion or mixing with waste water), sulphates (hardness, laxative effects), nitrates (lack of oxygen in blood, mathemoglobeanaemia) and phosphates (algal growth). Toxic chemicals like As, Ba, Cd, Cr^6, CN, Pb, Se, Hg, Zu, Cu and Ni cause damage to internal organs (kidney, liver, brain) of the human body. Some are carcinogenic. Based upon the waste water characteristics, undesirable effects due to water pollution are as follows:

(i) Soluble organics cause depletion of oxygen in water as they demand oxygen for their biochemical decomposition.
(ii) Suspended solids impair normal aquatic life as they may get decomposed and deposited at the bottom of water bodies.
(iii) Trace organics and phenols cause tastes and odours.
(iv) Heavy metals, cyanides and toxic substances are very harmful to living groups.
(v) Refractories resist biodegradation.
(vi) Colour and turbidity create unaesthetic conditions and may affect photosynthesis process.
(vii) Nitrogen and phosphorous stimulate and increase algal blooms which is undesirable.
(viii) Oil and floating matter are unsightly and retard reaeration of water.

(ix) Acids and alkalies affect fish and other aquatic life.
(x) Hot water discharges affect water solubility.
(xi) Inorganic materials may give rise to various problems such as hardness and corrosion.
(xii) Alkyl benzene sulphonate (detergents) leads to foam formation in water bodies.

Water pollution causes serious problems. It can be minimized by treating waste waters before disposal into water sources. Afterwards, water needs again purification before it is used for drinking purposes. Treatment of sewage, industrial wastes and drinking water have been briefly explained here.

Waste Water Treatment

General

Treated water, collects impurities after use in domestic sector, and becomes waste water. Sewage (domestic waste water) consists of faecal matter. Sullage means used water from baths, kitchen, etc. This waste water (99 per cent of water and 1 per cent solids) is transported through a network of sewers, away from residential units. Sewage contains human excreta that mainly causes water borne diseases. Organic matter decomposes and emanates foul gases and odours. Many impurities contaminate water sources and land surfaces.

Sewage contains (i) complex group of organic solids (carbohydrates, proteins, fats) which are biodegradable and (ii) suspended solids which can be removed in primary treatment. C, N, O, S, H, P from wastes utilize dissolved oxygen from water and undergo various chemical reactions with the help of micro-organisms (bacteria, algae, protozoa). The process of conversion of complex organic substances into very stable end products (as CO_2 and H_2O) is known as *decomposition of sewage*. Decomposition in the absence of dissolved oxygen converts sewage organics to CO, H_2S, CH_4, NH_3, organic acids and humus. Solids are partly converted into liquids and gases. Due to decomposition, strength and nuisance potential of wastes are reduced to an acceptable level.

The amount of oxygen required by the organic matter to undergo aerobic decomposition is known as *Biochemical Oxygen Demand* (BOD) at a given time and temperature. It is a measure of strength of waste. The amount of oxygen needed for oxidizing organic matter by chemical oxidation (without any biological action) is known as *Chemical Oxygen Demand* (COD). If the COD to BOD ratio is less than 1.5, waste water is more amenable to biological treatment.

Sewage contains different types of suspended, settleable and floating solids which are removed in settling tanks with scum removal devices. These solids are collected before biological decomposition as sludges. They require further treatment (stabilization) before disposal. The purpose of treatment is only to remove and stabilize the solid impurities and dispose the liquid wastes into water bodies or on land, safely.

Sometimes, industrial wastes cannot be effectively treated by biological treatment as they have extreme pH values, BOD to COD ratio less than 0.5, and more of toxic chemicals. Their strength can be reduced by equalization, neutralization, dilution with domestic sewage (after partial treatment of industrial wastes) or by segregating light and strong wastes.

Treatment Objectives

The tolerance limits for sewage and industrial effluents discharged into surface water sources are BOD_5 (20–30 mg/L); suspended solids (100 mg/L), nitrogen (50 mg/L), oil and grease (10 mg/L), phenol (1 mg/L), pH value (5.5 to 9) etc. Tolerance limits of waste water effluents to be discharged into the land for irrigation purposes are BOD_5 (200 mg/L), pH (5.5 to 9), total dissolved salts (200 mg/L), oil and grease (30 mg/L), chlorides (600 mg/L), sulphates (1,000 mg/L), etc. Sewage contains more harmful substances than are permissible for safe disposal. Hence treatment is essential.

The purposes of sewage treatment are as follows:

(i) To prevent water borne diseases
(ii) To remove organic solids from sewage and to convert solids into stable products by biological decomposition
(iii) To dispose treated effluents effectively and safely
(iv) To safeguard water sources from pollution
(v) To prevent nuisance and offensive odours

Schematic Outline

Primary treatment of sewage consists of screens, grit chamber, skimming tank, primary sedimentation and detritus tank. Secondary (biological) treatment consists of trickling filters, activated sludge treatment process, oxidation ponds, sludge digester, sludge drying beds, etc. (Figure 6.15). Advanced treatment is

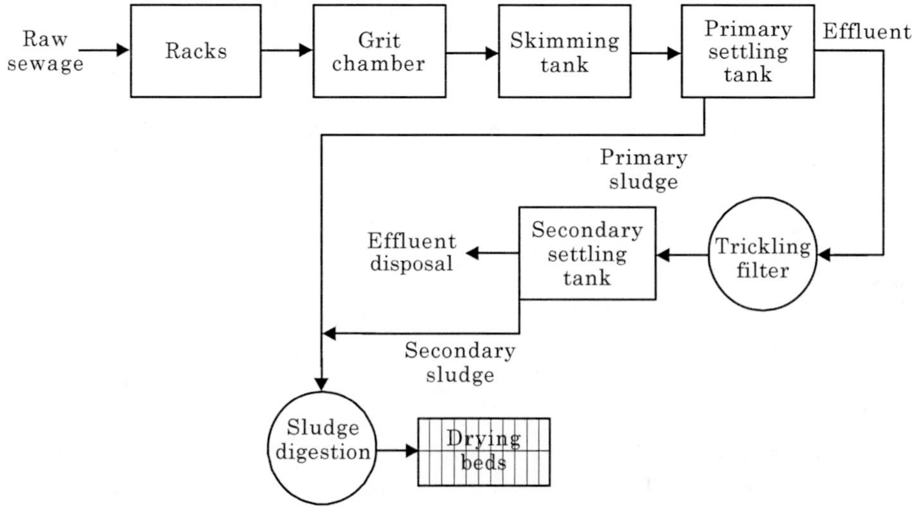

FIGURE 6.15 Waste water treatment plant layout.

given for the removal of specific characteristics from the sewage, as a special case. Chemical precipitation and chlorination are adopted only when they are required. Finally treated waste waters are disposed on land or water bodies.

- Screen chamber removes coarse materials, dead animals, logs of wood, and rags.
- Grit chamber removes heavy settleable inorganic solids such as sand.
- Settling tank separates (suspended, settleable) solids from liquids by gravity. [Figure 6.16(a) and (b)]
- Trickling filter (Figure 6.17) or activated sludge treatment plants (Figure 6.18) biologically decompose organic solids, reducing strength of wastes.
- Sludge digestion tank decomposes sludge anaerobically to produce methane and carbon dioxide and stabilize sludge.
- Drying beds ultimately separate solids from water.

Liquid effluents after treatment are disposed ultimately into nearby water bodies or waste lands depending upon the availability. A number of factors

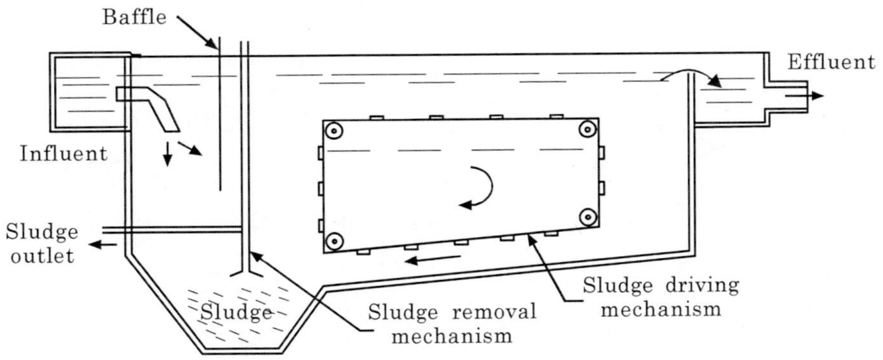

FIGURE 6.16(a) Rectangular settling tank.

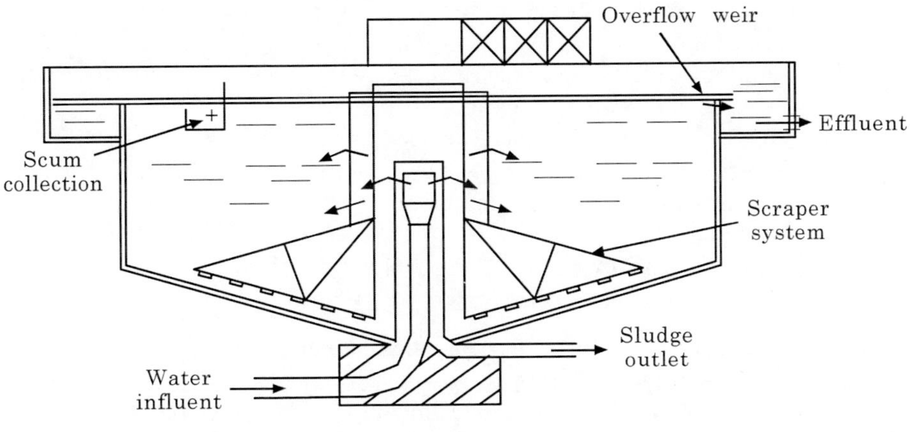

FIGURE 6.16(b) Circular sedimentation tank.

such as dilution, mixing, dispersion, settling, oxidation and reduction and help the river water in the self purification process. If treated effluents are discharged on land, crops can be grown with caution for gain.

Screens are provided to retain floating matter and coarse solids such as pieces of cloth, garbage and floating matter, so as to protect pumps and other units from clogging. A bypass channel is provided for the smooth passage of sewage flow, in case of clogging of screens. The screens may consist of parallel bars, rods, gratings, wire mesh or perforated plates. Screens may be kept ahead of grit chamber either in a separate chamber or in the same grit chamber.

Coarse screens (racks or bar screens) have an opening of about 50 mm or more whereas medium screens have 20 to 50 mm openings. Screens are inclined 30° to 60° to the direction of flow (to reduce velocity of flow and to increase the area of opening). Self cleaning velocity of 0.6 to 0.8 m/sec is maintained. Screens are fixed permanently and are cleaned manually or mechanically.

Grit chamber is provided to remove inorganic solids (specific gravity about 2.5, diameter less than 0.2 mm) like sand, ash and inert (non putrescible) materials, which may create operational problems. The simplest method of removing suspended impurities is by plain sedimentation.

As oils and greases are generally lighter than water, they are normally separated by natural flotation. In the skimming tank, the floating matter rises and remains on the surface of the waste water and the effluent flows out through deep outlet. Sometimes, a skimming trough may be provided in a sedimentation tank to collect the floating matter from sewage. Aeration easily separates scum from suspended solids, because of gentle agitation.

Detritus tanks (settling tanks) are provided to remove fine organic solids.

Trickling Filters

In these filters, (Figure 6.17) sewage trickles and percolates through filtering media. The essential components of a trickling filter are as follows:

(i) Rotary distributors, with two or four arms, supported at the centre. As the arms move, sewage passes into the bed uniformly on the entire surface area

(ii) Filtering media which consists of hard (granite) coarse broken stones 30 to 60 mm size and about 2 metres depth

(iii) Vitrified clay blocks which are used as under drains to collect filtered sewage

(iv) A ventilation shaft, for maintaining aerobic conditions inside the filter bed

(v) A dosing tank to maintain constant inflow into the filter unit which will also be useful for recirculation purposes.

The filter gives an opportunity for microorganisms to decompose organic matter from the sewage, utilizing the available oxygen from the atmosphere. They have been found to operate very efficiently, with a removal percentage of 70 per cent in BOD and suspended solids.

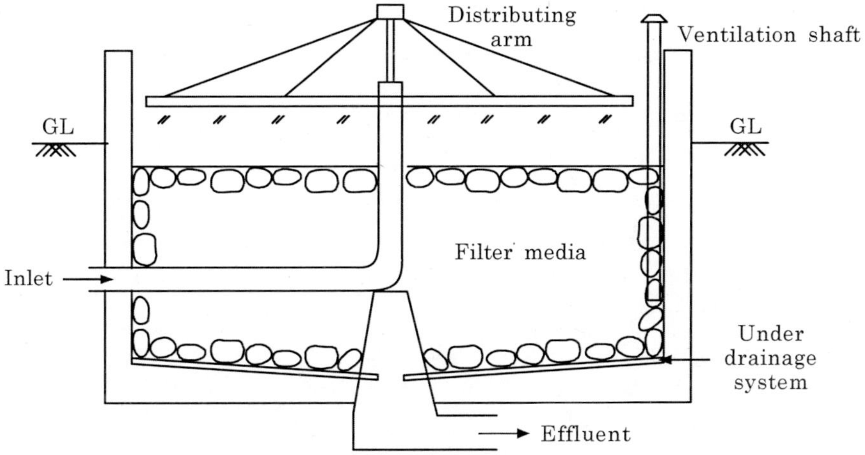

FIGURE 6.17 Trickling filter.

Activated Sludge Treatment

Sewage from a sedimentation tank enters into an aeration tank (Figure 6.18). Here 20 to 30 per cent of active sludge is mixed. The mixture is aerated and mixed in the tank for about 4 to 8 hours residence time. The microorganisms oxidise organic matter, in the presence of abundant quantity of oxygen in the aeration tank. Sewage is allowed to settle in a secondary sedimentation tank. This settled sludge has undergone aeration and has active microorganisms. So, some portion of this active sludge is recirculated into the aeration tank, where the mixed liquor with suspended solids is under continuous agitation. The excess sludge is disposed off from the secondary sedimentation tank. The treatment efficiency is more than 90 per cent.

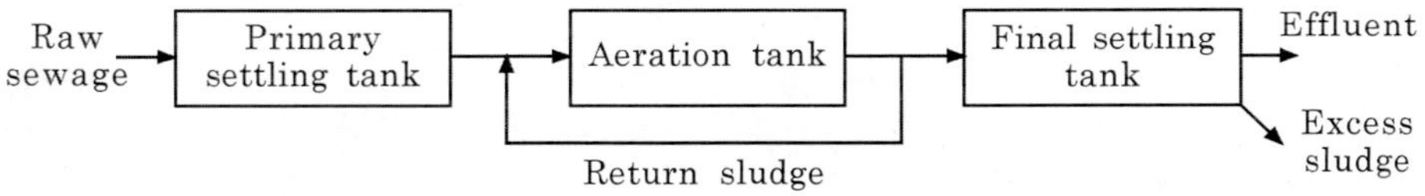

FIGURE 6.18 Activated sludge process.

Oxidation Ponds

These are simple, low cost holding basins 1–2 m deep taking some oxygen (air) from the atmosphere. Algae supply additional oxygen through photosynthesis process in shallow ponds, under aerobic conditions. If the depth is 3–5 m, decomposition takes place under anaerobic conditions. These ponds may accept higher organic loadings (Figure 6.19).

In facultative ponds, upper layers are under aerobic decomposition and lower layers are under anaerobic conditions. In the middle portion, algae may release oxygen to upper layers. From the settled organic load (benthal deposits), methane, ammonia, carbon dioxide, etc., may be released. These ponds may be kept in series for improved efficiency.

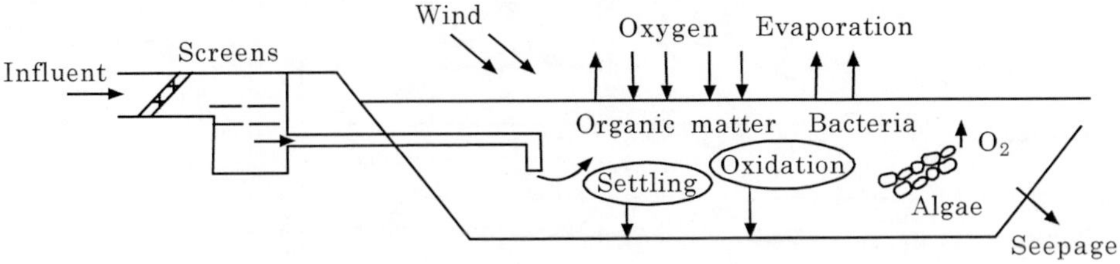

FIGURE 6.19 Oxidation pond.

Oxidation Ditch

This is an activated sludge treatment plant under extended aeration type, with detention time of 12 to 24 hours, and an air supply of 1.5–2.0 kg of O_2 per kg of applied BOD (Figure 6.20).

Oxidation ditches are water tight masonry chambers provided with horizontal axis rotors for agitating the sewage.

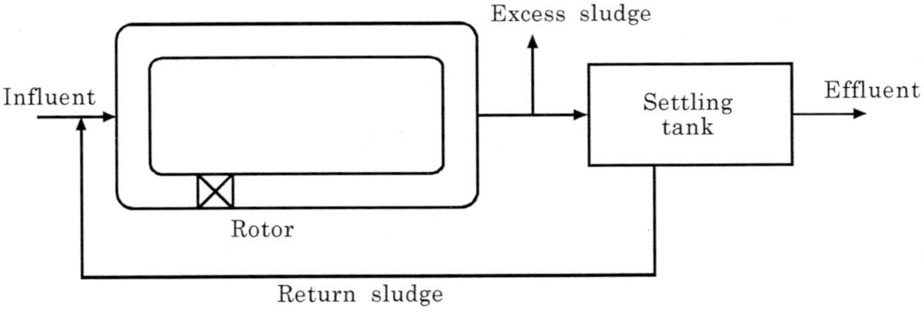

FIGURE 6.20 Oxidation ditch

RBC (Rotating Biological Disc Contractors)

For smaller installations, the microbial film is allowed to grow on the surface area of PVC discs rotating (partially submerged) in the direction of flow. These discs (2 m diametre and 100 mm thick) absorb (attach) organic load when submerged and takes oxygen when exposed to the atmosphere. Substrate utilization is a continuous process (see Figure 6.21).

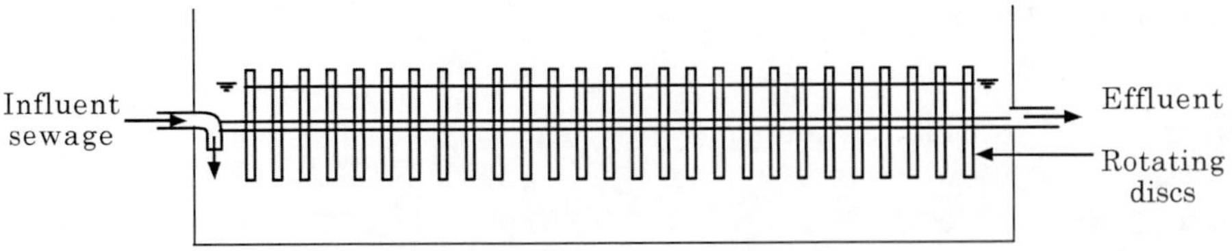

FIGURE 6.21 RBC.

Secondary Sedimentation Tank

After biological oxidation and nitrification, suspended solids settle quickly and more completely in a secondary sedimentation tank. The settled portion is sent to a digestion tank for further treatment.

Sludge Treatment and Disposal

Anaerobic treatment technologies are used for the effective treatment of municipal waste waters, sludge and industrial waste waters. This technology is particularly attractive as the energy required for operating the process is minimal compared to the energy required for aerobic processes. High rate digestor with mixing devices and temperature control, anaerobic filter, suspended and attached growth reactors, upflow anaerobic sludge blanket reactors (UASB) and fluidized bed reactors are developed and being used.

Sludge Characteristics

The main objective of primary sewage treatment is to separate solids from water. Solids (sludge) in the form of slurry contains putrescible substances which decompose further. Digested sludge dewaters quickly on drying.

Digestion of Sludge

Sludge is digested anaerobically, converting the waste solids into carbon dioxide and methane (Figure 6.22). Sludge volume is reduced as gases escape and lique-faction also takes place. Well digested sludge contains Nitrogen and Phosphates for use as a fertilizer. Methane gas is used as a fuel.

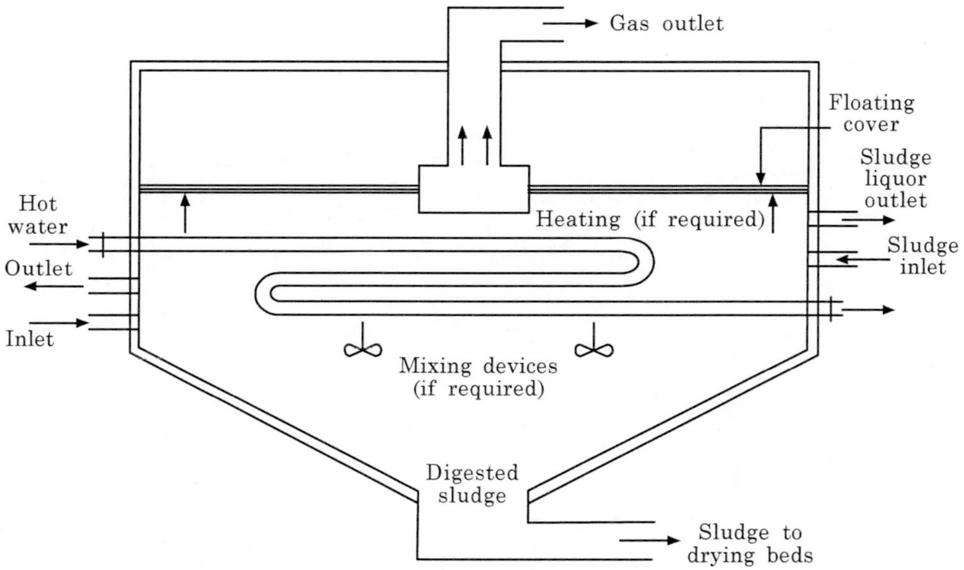

FIGURE 6.22 Sludge digestion tank.

Mechanism of digestion

Acid forming bacteria convert complex organic substances into simple organic acids. This acid production stage lowers the pH value. Methane forming bacteria converts the acids into methane and carbon dioxide. Decomposition of carbohydrates, fats and proteins will also take place. Since carbohydrates are digested easily, bacteria decompose them first. As the first stage of digestion nears completion (in 10–15 days), volatile organic acids and nitrogenous compounds are converted into H_2S, CH_4, NH_3, CO_2. Foaming occurs and the pH value also rises a little indicating the acid regression stage. More resistant proteins and organics are converted in the final stage resulting in alkaline nature of sludge. In this alkaline fermentation stage CH_4 and CO_2 are evolved.

Sludge Drying Beds

On drying beds water evaporates and percolates, leaving the solids to dry in 10–20 days time (Figure 6.23).

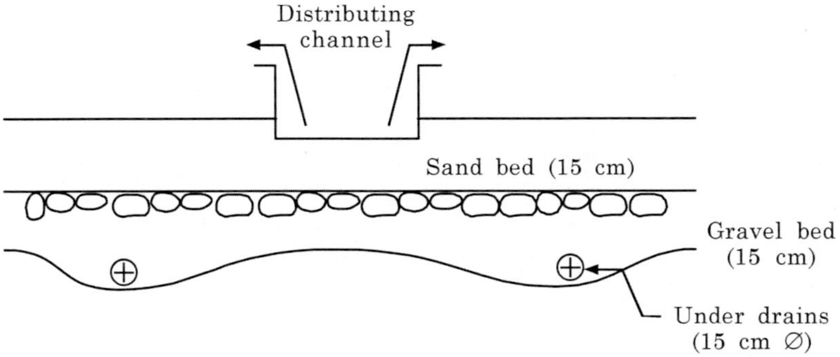

FIGURE 6.23 Sludge drying beds.

Septic Tank

In a septic tank (Figure 6.24) sewage and sludge are detained for 12 to 24 hours in a closed water tight chamber. Sedimentation and digestion are carried in this septic tank under anaerobic activity. Foul gases escape through a vent shaft.

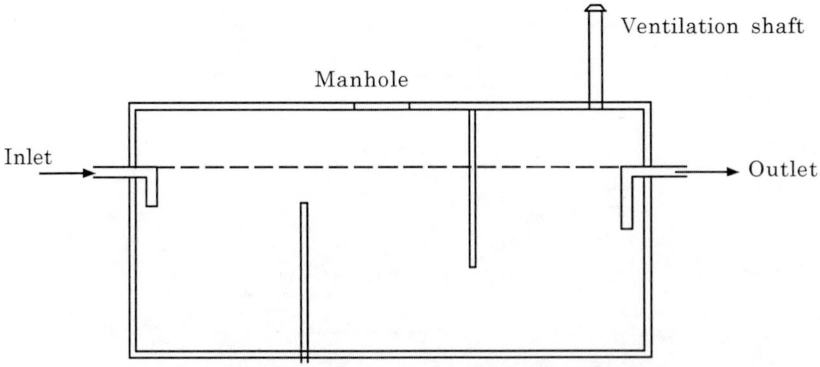

FIGURE 6.24 Septic tank.

The digested sludge is periodically removed once in a year or two, and disposed. The effluent from the septic tank is highly offensive and hence disposed into the ground through soak pits (Figure 6.25) or dispersion trenches. Septic tanks are highly useful for individual houses, hostels and small groups of populations.

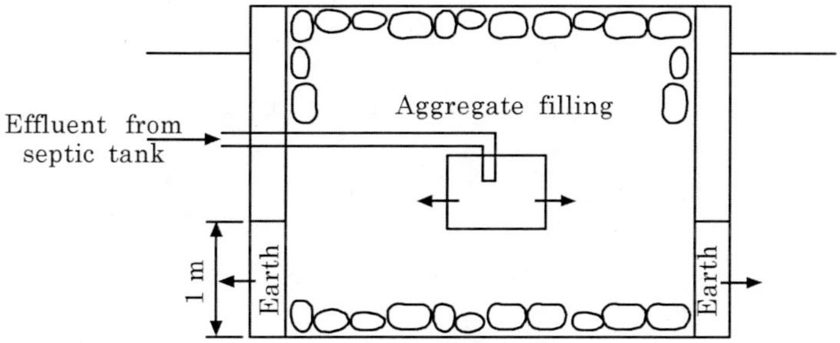

FIGURE 6.25 Soak pit.

Ultimate Disposal of Sewage

Disposal into a river. When organic wastes are added to river, suspended matter either tends to settle or is carried away downstream depending upon the hydrography of stream conditions. Aerobic bacteria utilize the pollutant material, in the presence of sunlight and oxygen available in surface waters and decompose organic matter. BOD of wastes is satisfied in course of time and flow distance in a river. River waters absorb oxygen from the atmosphere continuously. Due to dilution, sewage strength is reduced and due to natural reaeration, river waters are purified. Waters may not be fit for drinking purposes in some reaches of rivers where pollution is significant.

Sewage effluents discharged into surface water sources should not have more than 20–30 mg/L of BOD and 30 to 50 mg/L of suspended solids.

Marine waters have less dissolved oxygen, low settling capacity, high density, more tidal currents and wind action, and cause foam formation. But sewage may be disposed 1 or 2 km deep into sea.

Disposal on land. Waste water treated effluents are discharged on land for irrigation purposes if BOD (200 mg/L), pH Value (5.5 to 9), T.D.S. (1,500 mg/L), chlorides (500 mg/L), and sulphates (1,000 mg/L) are within tolerance limits. While successfully disposing sewage, sewage farming encourages growing of crops (which are not eaten raw and not in contact with sewage directly), such as grass, cotton, sugarcane, papaya.

Water application rate depends upon crop requirements, climate, soil characteristics, etc. Normally 100–200 m^3/d/hectare may be applied, whereas BOD load can be 50–60 kg/d/ha.

When sewage is applied continuously on land, soil gets clogged with sewage solids and land will not be in a position to receive further load. This sickness of soil is prevented by primary treatment of sewage, resting of land, rotation of crops, ploughing the land, improving soil drainage.

Water Purification

River water should be treated before using for drinking purposes to satisfy drinking water standards.

Purification of water is required for the following purposes:

(i) Safety for preventing water borne diseases
(ii) Palatability (mineral content, temperature) from aesthetic considerations
(iii) Wholesomeness of water taking the dissolved salts into account (in desirable concentrations)
(iv) Ensuring water quality so as to attain the permissible standards till the water reaches the consumer
(v) To remove turbidity, colour, taste and odour which causes the water to be distasteful
(vi) Supplying adequate quantities of water for various domestic purposes

- Sedimentation process removes suspended solids from water.
- Coagulation process (the addition of alum and precipitation in a settling tank) removes fine colloidal particles.
- Sand filter removes suspended impurities and bacteria.
- Chlorination destroys harmful bacteria as a final method of water purification before water is sent to the consumers. A residual chlorine of 0.1 mg/L in drinking water is acceptable.
- Aeration, ion exchange, adsorption, reverse osmosis, electrodialysis and a number of other methods are also available for water and waste water treatment.

A conventional treatment plant consists of three main units, viz., sedimentation (with or without coagulation), filtration and disinfection. Dissolved salts are removed only when necessary.

Sedimentation

The simplest method of removing suspended impurities is plain sedimentation. The water is allowed to stand quiescent or move very slowly until the suspended impurities settle to the bottom and relatively clear water is obtained.

It is a process of settling of particles, heavier than water, due to gravitational forces under calm and quiescent conditions. Particles settle as discrete (single, unattached) particles in a plain sedimentation tank. Particles combine together (coalesce, flocculate) and settle quickly in coagulation—sedimentation tank.

Two forces are acting on a particle. A downward pull or a driving (impelling) force and a frictional resistance, i.e., drag force, acting upward on the particle. The resultant force decides whether a particle settles, floats or is carried away with flow.

In a continuous flow sedimentation tank settling efficiency depends upon (i) size, shape and specific gravity of particles, (ii) temperature, viscosity and density of water, (iii) water flow velocity, (iv) disturbance from inlets and outlets and calm conditions (refer to Figure 6.16).

Tube Settlers

Tube settlers (Figure 6.26) are a bunch of plastic tubes 5 cm × 5 cm, 1 m long, kept in an inclined position as in a settling tank. Water flows upward and the settled material moves downward.

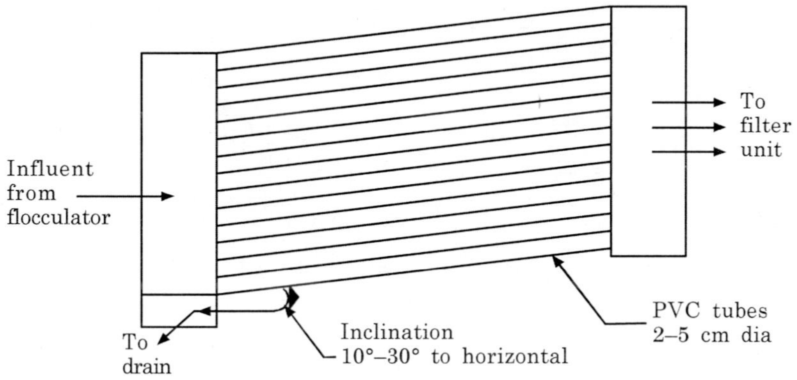

FIGURE 6.26 Tube settlers.

Coagulation

Very small particles of 1–100 m size (inorganic or organic), causing turbidity do not settle due to gravitational forces only. These colloids are minute, charged particles, have large surface area, exhibit Brownian (erratic, zig-zag) movement and are held in suspension. Colloids are stable. If the charge or potential is removed and smaller particles are made to combine into a bigger mass (by the addition of a chemical coagulant like alum or ferric salt), settling will take place quickly, easily and completely in a sedimentation tank. Process of coagulation is in three stages (Figure 6.27):

1. *Destabilizing.* Alum forms aluminium hydroxide (precipitate) in alkaline waters. These $Al(OH)_3$ flocs, i.e., positively charged trivalent Al ions, neutralize the negatively charged colloidal particles. This requires mixing of coagulant thoroughly with impurities. Coagulation of turbid water consists of a series of operations that result in the destabilization of the colloids leading to its removal from the suspending fluid.

2. *Aggregation and binding.* After the negative charges of turbid particles are neutralized (destabilized) by the positively charged $Al(OH)_3$, attraction causes aggregation of colloids. These minute flocs entangle some more impurities during floc formation stage, movement and descent, and settle down. Before finally settling, they collect maximum amount of turbid particles from water. So, flocculation is a slow mixing second stage of coagulation process, by which flocs form

and combine with impurities into big masses. This is achieved by slow stirring devices. It is to agglomerate destabilized particles under gentle agitation.

3. *Settling.* It is a final phase of coagulation process where finer suspended particles formed earlier into a big mass with greater affinity are removed in a settling tank or a clarifier.

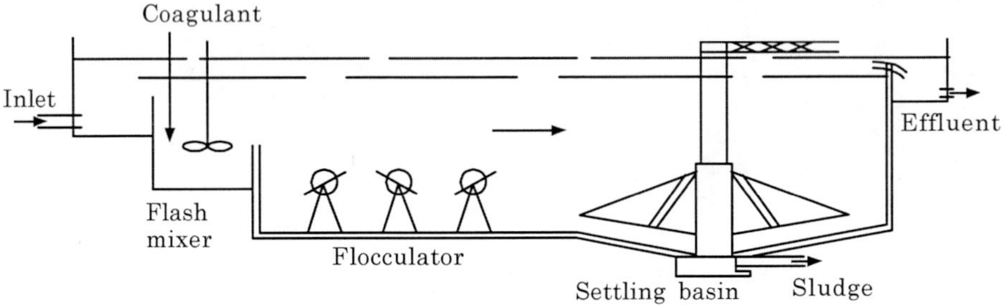

FIGURE 6.27 Clariflocculator.

Alum and iron salts (ferrous sulphate, ferric chloride or a combination) are used as coagulants. Iron salts produce heavy, dense, quick settling flocs in a wider range of pH. But alum is preferred as it is highly efficient in the normal pH range of water and is quite economical.

Filtration

Water with suspended impurities and bacteria passes through a bed of clean sand and the impurities are retained in the filter layers.

Mechanism

Pores between sand particles act like a sieve and retain bigger particles. Pores act as a miniature settling tanks. Due to physical and electrochemical forces, surface of the sand particles in the filter bed attract and attach (adsorb) turbid particles.

Bacteria utilizes (decompose) organic material deposited in the top layers of sand bed and forms a fine (slimy) layer. This layer is very effective in removing finer solids also. More deposition in the top layers attract more bacteria. So bacteria and solids are retained at the top layers only if the flow rate is low. Filters are stopped and cleaned when the loss of head reaches a maximum permissible limit or when impurities are coming into the filtrate.

Types of Filters

In gravity flow filters, two flow rates are adopted, one is a low rate of filtration (100–150 L/hr/m^2) known as *slow sand filter*, the other is a high rate of filtration (5000–6000 L/hr/m^2) known as *rapid sand filter*. Pressure filter maintains high rate (more than 8000 L/hr/m^2) (see Figure 6.28, 6.29 and 6.30). Diatomaceous filters, dual media (sand and coal) filters and upflow filters are also adopted.

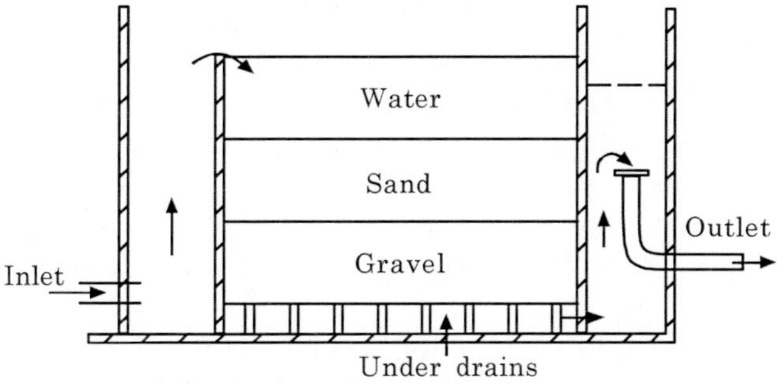

FIGURE 6.28 Slow sand filter.

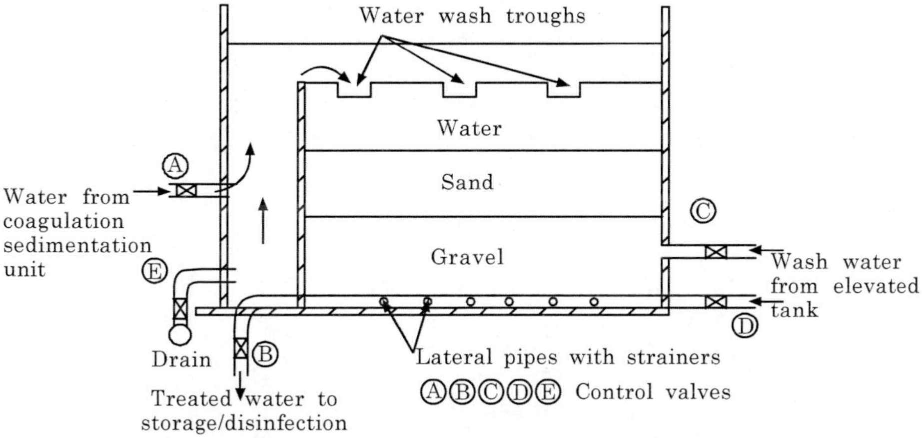

FIGURE 6.29 Rapid sand filter section.

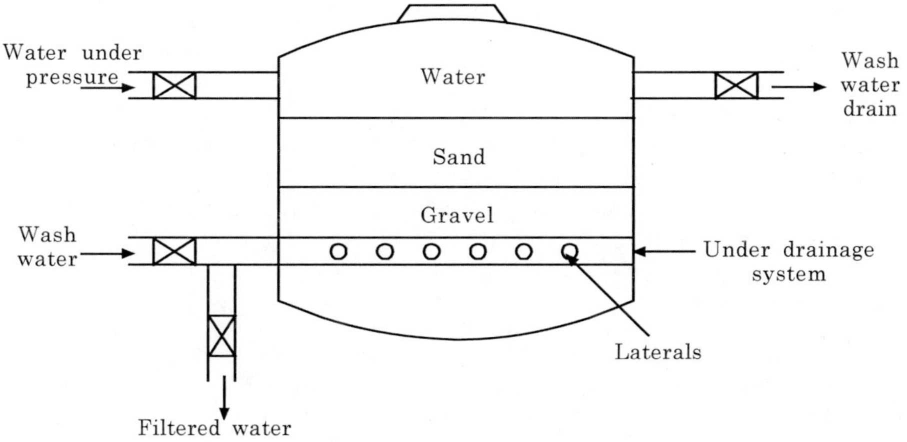

FIGURE 6.30 Pressure filter.

Disinfection

The main aim of disinfection is to destroy pathogenic (disease causing) bacteria and thus prevent water borne diseases. Chlorine is used as a disinfectant, normally after water filtration. Addition of chlorine to water results (hydrolysis) in the formation of HOCl and HCl and later HOCl ionises into H and OCl (free available chlorine). Chlorine combines with ammonia (chloramines), resulting in combined available chlorine. As pH of the solution increases, more HOCl dissociates into OCl ion. When pH value is less than 5, only elemental chlorine exists. Above pH of 10, mostly OCl remains. Hypochlorine is more powerful in the disinfection process. Chlorine penetrates into the cell wall and inactivates the bacterial cell metabolism.

Chlorination is done in the solid state (addition of $CaOCl_2$, bleaching powder) in small scale or liquid (or gas in solution) connected to water mains. It is always preferable to keep a residual chlorine amount of 0.2 to 0.6 mg/L to prevent further contamination in the distribution system.

Heavily contaminated waters need super dose of (excess) chlorination, i.e., about 5–10 mg/L. Dechlorination is to be followed later by the addition of SO_2 or $Na_2S_2O_3$.

The requirements of a good disinfectant are as follows:

(i) It should not be organism specific.
(ii) It should not render water toxic.
(iii) It should not demand specific conditions.
(iv) It should stay as a residual amount to prevent further contamination in distribution mains.
(v) Its action should be rapid and reliable.
(vi) It should afford feasibility in application.
(vii) It should be cheap and easily measurable.

Aeration

Aeration of water supplies removes bad tastes and odours and freshens water with oxygen.

In cascade type of aeration, water flows as thin sheets on horizontal steps. In spray nozzles, water droplets sprinkle into the air absorbing oxygen. In diffused aeration system, compressed air is sent into water tanks through diffuser pipes.

In all these simple devices, surface area of exposure is increased so that more opportunity is given for oxygen absorption. See Figure 6.31.

Ion Exchange Process

In these units, ions are transferred from liquid phase to solid phase and solid phase to liquid phase through cation or anion exchange resins. This is a reversible reaction and the chemical is regenerated for reuse.

Softening Process

Temporary hardness is removed by boiling (converting bicarbonates into carbonates for precipitation) or the addition of lime with alkalinity adjustment if required.

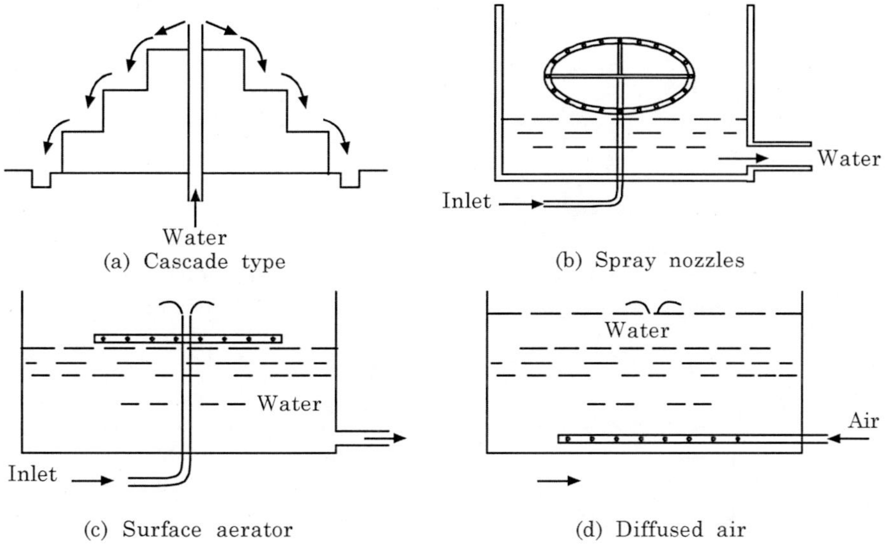

FIGURE 6.31 Aeration units.

$$Ca(HCO_3)_2 + Ca(OH)_2 \rightarrow 2CaCO_3 + 2H_2O$$

Permanent hardness is removed by lime soda process.

$$CaSO_4 + Na_2CO_3 \rightarrow CaCO_3 + Na_2SO_4$$

$$MgCl_2 + Ca(OH)_2 \rightarrow Mg(OH)_2 + CaCl_2$$

$$CaCl_2 + Na_2CO_3 \rightarrow CaCO_3 + 2NaCl$$

Zeolites (Na_2Z or Na-Ca-Al-Silicate) are used for the removal of hardness efficiently and completely.

$$Na_2Z + Ca \rightarrow CaZ + 2Na$$

Adsorption

A number of water pollutants diffuse onto the surface of a solid adsorbent. Organic chemicals (imparting colour, tastes/odours) heavy metals, fluorides etc., are removed using solid adsorbents like activated carbon, activated alumina, silica gel etc.

Activated carbon is inexpensive, non-selective, has a wide range and large quantities can be treated efficiently.

Membrane Process

Dissolved salts are removed in reverse osmosis and electrodialysis units. Potential of two solutions is different because of salt concentration separated by a membrane.

Reverse Osmosis

Liquid flows from a concentrated solution to a dilute solution across a semipermeable membrane of polymeric material (due to application of an external pressure or a driving force) which allows only water to pass through and not salts from solution.

Electrodialysis

When water flows through an array of membranes (cationic and anionic), various ions migrate towards electrodes, leaving behind pure water.

Corrosion control

Corrosion means deposition of substances on the pipe material, and eating away the material, resulting in the loss and reduction of the life of pipe. Acid waters are more corrosive. Corrosion is controlled by (i) removal of CO_2 from water, (ii) adjustment of pH, (iii) application of protective coatings on pipes, (iv) insulation and cathodic protection.

Others

Fluorides are removed by (i) the addition of alum and lime followed by clarification, (ii) passing through beds of activated alumina, (iii) ion exchange process through special resins.

Salinity is removed by the distillation (evaporation) and condensation of water vapour, electrodialysis, reverse osmosis and ion exchange processes.

6.3 SOLID WASTES DISPOSAL

Solid wastes are generated from various domestic, commercial and industrial activities which are classified as municipal wastes, industrial wastes and hazardous wastes and their constituents may vary from place to place.

Municipal Wastes

There are three main components of refuse—garbage, rubbish and ashes.

Garbage (or food wastes) consists of putrescible or decomposing organic matter such as vegetable, fruits, food material. They may be 0.10 kg/head/day on an average. These wastes are best suited for composting or biodegradation.

Rubbish consists of combustible and non-combustible solid wastes (at average per capita contribution of 0.25 kg/day), excluding putrescible materials. They include glass, metal, leather, textiles, paper, plastics, dirt and building materials used or demolished from construction industry. Ashes and residue products are the remainings after combustion.

Sometimes wastes from street sweeping, dead animals and animal dung may join with solid wastes. It is a difficult task to separate various constituents from a normal mixed refuse of a dust bin. But segregation is practiced to recover some useful material. Plastic bags and material may be separated as they are not bio-degradable. Magnetic material, paper, card board, metals, and glass bottles may have a resale value in the market.

Characteristics of Solid Wastes

Quality and properties vary due to community habits, standard of living, place of collection, frequency of collection, season and climate. Solid wastes are analyzed for the purpose of deciding over the method of disposal and resource recovery systems.

Bulk volume or density is useful to know the compressibility of wastes. Moisture content and temperature indicate the qualities of putrescibility and bacterial activity. Organic content, carbon to nitrogen ratio and moisture content are useful parameters in the composting (biodegradation) process of disposal. The generation of energy is dependent upon the calorific value of the wastes. If inert matter is more, refuse can be easily disposed off in landfill operations. In case of recovery of materials from the wastes, economic feasibility is decided when the available quantity is known. Recovery and reuse of paper, plastic bags, oil containers, glass bottles etc. significantly reduces the load of solid wastes.

Methods of Disposal of Solid Wastes

Municipal wastes are disposed effectively by land fill, composting and incineration processes.

(i) *Land fill:* Solid wastes are placed in pits, trenches or low lying lands, compacted and covered with a layer of earth. In two to three years, solid waste volume shrinks by 20–30 per cent and the land can be used for parks. One cubic metre space can take more than 500 kg of refuse. This method is simple and refuse can be disposed without any segregation of constituents. The main problems associated with the method are availability of land, soil and ground water pollution. Site selection is based upon soil conditions and topography, surface and ground water hydrology, geological characteristics, travel distance and nuisance value to neighbouring areas. Percolation of leachates should not create any adverse impact on water resources.

(ii) *Incineration:* Burning of combustible refuse is the safest process. Incineration is quite effective at about 1000°C when the garbage content is more than 50 per cent of the total refuse content and calorific value is more than 1600 k cal/kg. Pyrolysis is the destruction of hydrocarbons at high temperatures to obtain energy. If needed, additional fuel or oxygen supply is provided for effective burning.

(iii) *Composting:* Solid wastes rich in putrescible organic matter is used for generation of biogas by anaerobic digestion process (in the

absence of oxygen). Gas is used for fuel or lighting purposes. Liquid portion (slurry) is a rich manure containing nitrogen and phosphates. Composting process needs about 30 per cent moisture content and carbon to nitrogen ratio of 30:1 in the waste matter. The pre-requisite for composting is the removal of non-biodegradable materials like plastics, metals, glass from solid wastes. Addition of earth worms help in bio-decomposition of wastes in the vermi composting process.

Solid wastes generated from various industrial sources are given below:

Industry	*Wastes*
Textile units	Fibre and cloth residues
Paper industry	Paper and fibre residues, chemicals
Petroleum refineries	Coal tar products
Leather industry	Dyes, oils and leather
Chemical processes	Organic and inorganic chemicals
Metal industry	Metal scrap

Hazardous Waste

Cyanide, heavy metals, oils and emulsions, phenols, asbestos, acid or alkaline slurries, pesticides are some of the toxic and hazardous products from industries which pose a potential danger to the human life. Biomedical wastes from laboratories and hospitals also create dangerous infections and incineration under controlled conditions is the best solution.

6.4 SOIL POLLUTION

Land is a resource with minerals, nutrients, oil products to support life on earth. Vegetation grows only if the top soil is fertile. Human beings derive food only from plants and crops. Formation of top soil is a slow process, replenishment takes considerable time. The following activities disturb the top layers of the soil:

(i) Construction and mining activities
(ii) Soil erosion by wind and water
(iii) Disposal of solid and liquid wastes
(iv) Use of chemicals (fertilizers and pesticides)
(v) Contamination with micro organisms
(vi) Waterlogging and salinity

Soil has a natural self purification property to some extent. Land acts as a filter in removing solid impurities from waste water. These impurities decompose into nutrients to enrich the soil. But continuous disposal of waste waters without adequate treatment pollutes the soil beyond recovery. Soil contamination leads to land sickness and finally loss of land for ever. These impurities also reach ground and surface water sources causing pollution.

Rivers carry rich sediments and deposit in reservoirs. Erosion at one place and deposition at another place occurs due to surface run off. Similarly wind action affects top soil. Transportation of top layers of soil result in the loss of nutrients and minerals. Excessive use of fertilizers and pesticides in agricultural operations and the residual chemicals remain in the top layers of soil. Toxic chemicals kill microorganisms favourable for plant growth. Toxic chemicals also reach drinking waters. Many harmful chemicals reach human beings through food (vegetables, fruits, fish, poultry, meat) products. Excess of Na, Mg, Ca, K, S, Zn, Fe inhibit plant growth and reduce crop yield. Compounds containing arsenic, mercury, chromium, nickel, lead, cadmium, are toxic to human life. Fluorides also affect the plant development. Organic wastes entering the soil pores, decompose and spread infections through pathogenic bacteria. Diseases caused by Hookworms and helminthus are common in villages due to soil pollution.

Some plants are very sensitive to soil acidity or alkalinity and salinity. So land becomes unfit for irrigation due to water logging.

Soil Remediation

Prevention and proper care saves the fertile soil. Soil, water and plant relationship should be understood. Mineral cycles involving nitrogen fixation, nitrification are to be protected to preserve soil fertility.

The following remedial measures are suggested:

(i) Treat waste water adequately before land disposal. Prevent entry of leachates into the soil.
(ii) Control tree plantation at proper places to prevent soil erosion.
(iii) Use natural fertilizers and biological agents for pest management. Preserve soil friendly bacteria and soil micro-nutrients.
(iv) Plough or mix the soil to improve aeration, porosity and permeability. Select or change the crop patterns suitably to replenish the nutrients in top soil.
(v) Remove water logging and salinity conditions.
(vi) Cultivate grass only on polluted lands.

6.5 MARINE POLLUTION

Domestic and industrial wastes reach ocean through river water. Solid and hazardous wastes are also disposed into oceans. Floating oil and solid matter reach the coastal shores and make surroundings unhygienic. Residual chemicals from agricultural fields also join sea. Industries located near the coastal line may dispose the effluents into sea. Another major pollution is due to oil spills in sea water. It may be due to oil drilling operations or due to leakage from marine shipping tankers. Floating oil forms a surface layer on the top of water surface which is detrimental to marine life. Aquatic fish life and organisms like phytoplankton are affected. Birds and animals living on sea food; plant life needing sunlight and fishermen are the victims. Plastics, lead, mercury affect the marine ecosystem. Tourism also affects coastal areas from the disposal of wastes. Harmful chemicals reach human beings through fish and aquatic life. Marine pollution can be controlled by the following ways:

(i) Treating all wastes before disposal into sea
(ii) Using chemicals to convert oil into solids or absorbents to contain spill oil
(iii) Breaking and degrading oil through microorganisms
(iv) Emulsification of oil and dispersal
(v) Skimming through ships
(vi) Proper coastal area management

6.6 THERMAL POLLUTION

Disposal of recycled water used in cooling plants of power steel nuclear and other industrial units into water bodies causes thermal pollution.

Water temperature affects the following:

(i) Food and oxygen requirements and their availability in water bodies
(ii) Microbial activity as species need optimum temperatures for their metabolism
(iii) Solubility and saturation levels
(iv) Chemical reactions
(v) Water properties like viscosity, density
(vi) Changes in ecosystems

The easiest solution is to keep the hot water in cooling tanks for some time to reduce water temperature and then dispose into water bodies. Recirculation to the extent possible may be a reasonable solution.

6.7 NOISE POLLUTION

Noise is unpleasant, unwanted, loud sound causing nuisance and disturbance to the receiver. Noise interferes with normal human activities. Continuous exposure to loud noise leads to physiological disorders or psychological stress. Irritation, frustration, fatigue, headache, anxiety, hypertension, insomnia, hearing loss, mental depression are generally associated with exposure to high intensity of noise for short durations or moderate levels for a prolonged period. Discomfort and lack of concentration affect worker's efficiency and even lead to accidents. Sudden explosions create cracks in building walls and even break window glasses.

Normal sound levels are indicated as follows:

Normal conversations inside offices, residences	—	50 to 60 dB
Hotels and commercial areas	—	60 to 70 dB
City traffic	—	70 to 80 dB
Foundry unit	—	90 to 100 dB
Air craft	—	100 to 110 dB
Blowers and drilling operations	—	Above 100 dB
Hammer fall	—	120 dB

It is observed that:

80–85 dB sound causes discomfort
100–110 dB sound causes stress
120–130 dB sound causes nausea, dizziness, and pain in the ear.

Higher levels cause permanent damage to the hearing mechanism.
All these effects of noise pollution depend upon the following factors:

(i) The distance of the observer from the source of noise generation
(ii) Intensity level and duration of exposure
(iii) Combined effect due to multiple sources producing different sounds

Acceptable noise levels are 50 dB in silence zone near hospitals and schools, 65 dB in commercial areas and 75–80 dB in the working environment.

Noise Control Measures

A number of provisions are made in the Factories Act (for industrial noises), Motor Vehicles Act (for automobiles) and Public Nuisance Act (for using loudspeakers in residential areas). People should rigidly follow the rules and regulations so that sound levels do not exceed the prescribed limits. Some measures to control the noise are given below:

(i) Plan traffic movement for a smooth flow;
(ii) Create silence zones near schools, hospitals, prayer halls and prevent the use of loudspeakers, crackers and vehicle horns;
(iii) Plant trees on road margins to absorb and dissipate sound energy levels;
(iv) Regulate time of exposure and sound levels in the industrial atmosphere; Workers should use personal protective ear plugs;
(v) Ensure smooth working of machines with proper repair and maintenance by tightening loose parts and using lubricants for machines;
(vi) Rigidly fix the machines on concrete blocks and place absorbent material to reduce industrial noise due to impact and vibration. Rubber, felt, cork and steel springs reduce transmission of mechanical vibrations. Asphalt material also dampens the sound levels; and
(vii) Construct acoustical cabin enclosures for operations like drilling, grinding, to arrest noise levels at source. Areas producing noise may be segregated to create a buffer zone.

SUMMARY

(i) Various industries from combustion and manufacturing processes release a number of pollutants (CO, SO_x, NO_x, HC) into the atmosphere to cause air pollution, which is harmful to human beings, plants, animal life and damage to materials. Particulate matter can be removed by settling chambers, wet scrubbers, bag filters, e.s.p. units, at source. Gaseous impurities can be removed by adopting absorption, adsorption and other suitable devices.

(ii) Various impurities (physical, chemical, bacteriological) from domestic sewage join drinking water sources and spread water borne diseases. Solids (organic, inorganic, suspended, dissolved), acids, alkalis, heavy metals and various toxic pollutants released from industries also reach drinking water sources. Fertilizers and pesticides used in agricultural operations, leachates from wastes also pollute water supplies. They render water unfit for drinking purposes. Dissolved oxygen is consumed for decomposition of organic matter and affects the aquatic life too. Waste water must be treated before disposal and water must be purified before domestic use. After primary treatment (sedimentation), waste water undergoes biological treatment and solids are digested before disposal, ultimately into water sources or on land. Drinking water needs sedimentation, filtration and disinfection treatment essentially.

(iii) Solid wastes have its nuisance value. Decomposing organic wastes are treated/disposed in composting units. Combustible matter is incinerated safely. Inert matter in general, is disposed in low lying areas, effectively. Soil fertility is degraded due to erosion, use of chemicals, disposal of wastes, salinity and water logging. When land is contaminated, harmful chemicals reach human beings through crop yields and vegetation. Noise (unwanted sound) levels disturb the peaceful atmosphere and lead to physiological disorders due to stress.

All kinds of pollution require adequate measures for the safety and well being of the mankind.

KEYWORDS

Air pollutants (sources, harmful effects, removal of particulate and gaseous impurities); Domestic and industrial effluents; Marine pollution; Noise pollution and control measures; Soil fertility, Pollution and remediation; Solid wastes (characteristics and methods of disposal); Wastewater, water pollution, water purification; Thermal pollution; Vehicular pollution.

SOLVED QUESTIONS

1. Give the chemical reactions leading to PAN formation.

$HC + O \rightarrow RCO$ (acetyl radical)

$RCO + O_2 \rightarrow RCO_3$ (Peroxy acetyl radical)

$RCO_3 + NO \rightarrow RCO_2 + NO_2$

$RCO_3 + O_2 \rightarrow RCO_2 + O_3$

$RCO_2 + NO \rightarrow RCO + NO_2$

$RCO_3 + NO_2 \rightarrow RCO_3\,NO_2$ (PAN)

(*Note:* Photochemical reactions need NO_x and HC in the atmosphere)

2. Write a note on the atmospheric stability and dispersion of air pollutants.

Horizontal movement of wind is responsible for the reduction of pollutant concentration with respect to time and distance from the source of emission. If a rising parcel of air remains warmer than the surrounding air, or if a descending parcel remains cooler, then the displaced air quantity continually moves away favouring dilution and dispersion. This is an unstable condition in the atmosphere. Otherwise in a stable condition, air movement is restricted and pollutants concentrate at a place. Air stability is decided by the temperature gradient in the vertical direction in the prevailing atmosphere.

Inversion is an unfavourable situation for dispersion of pollutants when temperature increases with altitude.

3. What happens when waste water is disposed into a river?

- Suspended matter settles or is carried away by the downstream.
- Aerobic bacteria decompose pollutant material in the presence of oxygen available in the river water.
- River water absorbs oxygen from the atmosphere.
- Dilution reduces the strength of waste matter.
- Polluted water is unfit for drinking purposes.
- Oxygen level is a measure of pollution level and essential for fish life.

4. Explain the utility of a septic tank.

Septic tank (Figure 6.24) is a closed water tight chamber used for the disposal of wastes in individual houses or hostels not covered by under drainage system. Waste solids settle in the tank. Anaerobic decomposition takes place releasing foul gases. The effluents from septic tank are highly offensive and hence should be disposed through soak pits or dispersion trenches.

5. What are the methods available for the removal of dissolved inorganic impurities from waste water?

- Ion exchange through resins
- Dialysis through semi-permeable membranes
- Reverse osmosis through difference in pressure
- Evaporation to concentrate mineral salts

6. Name some organic pollutants.

Insecticides like Aldrin, Chlordane, Dioldrine, Endrin, Heptachlor, Polychlorinated Bi Phenyls (PCB), Dioxins, Furans.

7. Explain the role of microorganisms in decomposing matter.

Microbial population is a mixed, heterogeneous group consisting of bacteria, fungi, protozoa, rotifers etc. They degrade organics (soluble, suspended or colloidal) into stable end products which do not cause pollution. Microbes use waste products as food for their respiration and synthesis to continue their growth and activity. They will decompose matter in the presence (aerobic) or absence (anaerobic) of oxygen. Final end products are safe to join environmental sources.

8. List some waste products from electronic industry.

Toxic chemicals and heavy metals like lead, cadmium, mercury, barium etc. from computers, television sets, cell phones (casings, circuit boards, switches, cathode ray tube, screens) to an extent of millions of tonnes of waste, are deposited into the environment. Some of these substances are not bio-degradable.

REVIEW QUESTIONS

1. List out the pollutants causing air pollution.
2. Mention the sources responsible for air pollution.
3. State the harmful effects of any three chemicals on human beings, plants and materials.
4. Name the devices for the removal of particulate impurities.
5. Write short notes on:
   (i) Automobile exhaust
   (ii) Indoor air pollution
   (iii) Secondary pollutants
6. List out the various impurities associated with water quality.
7. What are the characteristics of domestic wastes?
8. Mention the harmful effects if waste water is disposed into a river without treatment.
9. Explain the measures adopted to prevent soil pollution.
10. Indicate suitable methods to reduce noise pollution.

11. Name some pollutants affecting your area.
12. What are the effects on environment due to resource extraction?
13. What is zero decibel?

OBJECTIVE TYPE QUESTIONS

Fill in the blanks with suitable words:

1. One example for air pollution due to natural sources, is ________.
2. ____________ is a major source of air pollution.
3. ____________ reduces oxygen carrying capacity of blood.
4. Permissible level of SO_2 in ambient atmosphere is ____________ $\mu g/m^3$.
5. Maximum permissible concentration of nitrates in drinking water is _________ mg/L.
6. Possibility of sewage contamination is indicated by the presence _________ in drinking water.
7. Disinfection is a process of ______________.
8. ______________ is needed for the decomposition of organic matter.
9. Absence of fish along a river indicates ____________.
10. Suspended solids in water are measured by _________.
11. Wastes from _____________ industry is likely to have high BOD.
12. ____________ is a toxic inorganic chemical.
13. Fuel adulteration and impurities in fuel, cause ________ pollution.
14. Sea breeze occurs in coastal areas during __________.
15. CO_2 production in India is about __________ whereas it is 20 MT/head/year in advanced countries.
16. Gas chromatography is used for detecting __________ in a sample.
17. Activated carbon is used for controlling ___________ pollution.
18. Alum is used in _________ treatment process.

CHAPTER 7

ENVIRONMENTAL CONCERNS

TOPICS COVERED

- Urbanization
- Industrialization
- Agricultural Revolution
- Inversion
- Acid Rain
- Greenhouse Effect
- Ozone Layer Depletion
- Global Warming
- Hazardous Wastes
- Disaster Management

7.1 URBANIZATION

25–30 per cent population of the country live in crowded and densely populated cities and major towns. This is mainly due to more employment opportunities, educational, medicare, and infrastructural facilities. Migration from rural areas to urban towns and cities is a regular and continuing phenomena. An important observation is that 25–30 per cent of city's population live in slum areas without adequate basic living conditions. Some times, city limits expand and encroach the industrial zones, water bodies, agricultural lands, forest areas and even mountain slopes. Probably, Master Plans are not properly implemented or planning is not commensurate with population growth or all rules and regulations are violated. Whenever there is rapid increase in population growth, due to some special reasons like setting up a steel plant near the town, it is very difficult to create sufficient infrastructure within a short interval of time. During the last 30 years, agricultural land decreased by 30 per cent due to urbanization.

Problems Associated with Urbanization

There will be heavy demand for housing. Individual houses in 20 to 30 sq. m. of land have given place to individual flats of 2–3 sq. m. land area per head, A 15 cm. water main catering to the need of 100 houses in a street, has to distribute the water supply to 1,500 houses in the heart of city. Vehicles moving on the street have increased by 10 times or more. Commerce and other service requirements are needed additionally. Then comes the development of housing colonies on the outskirts of a major city or town. This demands transport facilities to city, increasing the traffic congestion on the roads. Whenever new housing colonies are planned, residential and commercial units are not provided for the service sector initially which leads to slums in due course.

Each and every dwelling unit requires water supply and drainage facilities. Normally 200 to 300 litres of water is needed for a single person per day. It may not be possible to supply the required quantity of potable water. Only few hours of water supply may be possible when the resources are limited. Waste water collection, treatment and disposal facilities are available, only in some developed portions of the cities. In medium towns, drainage system does not exist. There is no provision for sewage water disposal. In the city slums, individual houses do not have proper latrines and not even one public tap or bore well water supply system is provided for hundreds of slum dwellers.

(i) When more vehicles are on the road, traffic congestion and pollution from gaseous emissions are common.

(ii) Consumers life style is different in urban conglomerations. Solid wastes coming from each housing unit is more and varied. Most of the refuse quantities are either burnt out or disposed in low-lying areas, which leads to unhygienic and unsanitary conditions.

(iii) As the built up land area is going to be more, less open space is available for ventilation and vegetation. This results in an increase in temperature of the urban atmosphere. Ground water depletion and disappearance of surface water bodies add to the problem of drinking water supply.

(iv) Quality of life in the urban environment appears to be more strenuous due to more work pressures and less comfort levels for an average citizen.

Carrying Capacity of a Region

It is possible to understand the trends of a city growth in the last one or two decades from the available records. Population densities in various localities, expansion of city limits due to new colonies, traffic congestion and accidents rate, availability of educational facilities, hotels, hospitals, parks, theatres etc.

may be regarded as a part of growth. An observation on the construction activity, land use patterns, ground water table, quality of water bodies, location of industries, number of vehicles on the road, power supply demands and other activities help in preparing a base map to indicate the present status of any city. This map should also identify the difficulties due to pollution of water resources, pollution due to industries, pollution from vehicles, nuisance from solid waste disposal methods, heavy demand for energy resources, noise pollution and slum locations. All these details always depend upon the standard of living and quality of life assured to the population in that specific area. Now the stage for future growth of a city can be planned with the proper inputs in the mind. An action plan can be prepared depending upon the future requirements and carrying capacity of that particular region. This type of planned growth will not create stress on the natural resources and infrastructural facilities available. So, special care has to be taken during expansion of city limits. The basic needs are to be satisfied by providing protected water supply, effective drainage collection system, educational institutions, medical facilities, good roads with sufficient width and adequate greenery. It means there is an upper limit for growth of an urban region where the sustainability is in doubt.

Remedies

A number of plans may appear on paper but policy implementation is difficult. Some remedial measures are suggested below:

(i) Effective method of reducing the migration trends to urban areas, is to understand clearly the reasons for the migration. Education, unemployment, social services and commercial activities may be uniformly spread to other locations, rather than concentrating in major cities. Some of the headquarters of government departments may be shifted to different towns as communication facilities are possible and available in all respects from the capital city.

(ii) Development of satellite townships around major urban centres, with all infrastructural facilities should be encouraged, where planned growth is possible. This will reduce pressure on the heart of the city.

(iii) City's master plan should clearly specify the various zones for residential and industrial activities without disturbing vegetation and water bodies.

(iv) Slum areas also need the same basic facilities of living space, drinking water and drainage facilities as any other developed localities. An integrated development of slums in a society is a big asset for all.

(v) To reduce the demand on power supply, alternate energy can be obtained from the solid wastes (decomposition/burning) or from solar, wind and other sources. As urbanization appears to be inevitable to some extent, care should be taken to provide the necessary infrastructural facilities for the growing population. Care should be taken to prevent, stop and control all sorts of pollution from all sources of emissions. Urban growth has its own limitations and checks.

Rural areas may be developed into model townships with all basic amenities required for quality living.

7.2 INDUSTRIALIZATION

Solid, liquid and gaseous waste products are released into the atmosphere from various processing industries. Strength and volume of these wastes are known. They vary vastly from industry to industry depending upon the raw materials used and the manufacturing process. Domestic wastes have a neutral pH and exert an oxygen demand of 200 mg/L. But the liquid effluents from an industry may be extremely acidic or alkaline in nature. The oxygen requirement may also be 10 to 100 times even, in comparison with sewage. Hence an industry can do much damage if the effluents are not properly treated before disposal.

Characteristics of liquid wastes from selected industries are given below:

(i) Dairy plant—BOD = 1,000 mg/L, solids = 1,000 mg/L, oil and grease, odours, putrescibility
(ii) Distillery unit—Dark brown colour, solids = 50,000 mg/L, nitrogen, iron, sulphates, chlorides and phosphates. BOD = 40,000 mg/L
(iii) Fertilizer plant—Arsenic, fluorides, phosphates, chlorides, sulphates and nitrogen compounds
(iv) Textile unit—Highly alkaline, colour, solids, Ca, Mg, Na, Zn and sulphates
(v) Paper plant—Alkaline, brown colour, fibrous matter, lignin and caustic soda
(vi) Pharmaceutical unit—BOD = 5,000 mg/L, extreme pH range, more solids, toxic organics, sulphides, phosphates, strong odours
(vii) Tanneries—BOD = 3,000 mg/L, highly alkaline, solids = 10,000 mg/L strong odours, colour, organics, sulphides, chromium

Characteristics of gaseous impurities from selected industries are given below:

(i) Paper plant—Mercaptons, dusts, SO_2, H_2S
(ii) Tanneries—Sulphides and mercaptons
(iii) Metallurgical units—Dusts, fumes and oxides
(iv) Chemical industry—Acid fumes
(v) Fertilizer unit—NO_x, HF, NH_3, Dusts of coal and sulphur
(vi) Thermal power plant—Flyash, NO_x, SO_2, HC, CO
(vii) Petrochemicals—Particulates, dusts, hydrocarbons, NH_3, NO_2, CO

Characteristics of solid wastes generated from selected industries are given below:

(i) Iron and steel—Process dusts, sludge from recovery units

(ii) Petrochemical units—Organic residues, oily sludge, residues from various units
(iii) Paper unit—Used chemicals, highly corrosive sludge
(iv) Textiles—Chemical residues
(v) Tanneries—Chromium bearing sludge
(vi) Electronic industry—Lead, cadmium, barium, mercury, plastic compounds

After independence, visionaries and great leaders have given priority for rapid industrialization for the development of the country. 'Industrialize or perish' was the slogan then. People have now realized that progress of the nation is associated with the depletion and degradation of natural resources as some of the industrial pollutants are indiscriminately disposed without adequate treatment.

Remedies

(i) Industries should completely treat all types of generated wastes as per the prescribed standards before disposal outside their factory limits.
(ii) If necessary, polluting industries should be closed or shifted far away from the human environment. Care should be taken at the time of their location.
(iii) Law should take its own course while controlling the pollution levels.

7.3 AGRICULTURAL REVOLUTION

Along with industrialization, priority was also given to the agricultural sector so as to cater to the needs of ever increasing population numbers. This has resulted in the construction of large dams and fertilizer plants to bring more lands under cultivation and increase the crop yield. However, number of side effects have been noticed all over the country and lot of agitations are going on against 'big dams'. The following observations are noteworthy:

(i) Storage reservoirs are leading to water logging which is detrimental to the fertility of soil.
(ii) Lands, forests and villages are getting submerged due to the construction of water retaining structures. Human rehabilitation has become a major problem.
(iii) Fertilizers and pesticides used in agricultural fields are joining water bodies causing health hazards to human beings.

The solution lies in the proper selection of water resource projects and use of natural fertilizers and bio-pesticides.

7.4 INVERSION

Atmospheric temperature normally decreases with altitude. A certain mass of air rises and moves up vertically when it is warm until its own density equals with its surroundings. Then it becomes lighter and cooler. Warm air expands, and moves to adjacent regions of lower temperature. Cooler air mass gets compressed, becomes heavier and descends down. Changes in the internal energy result in temperature variations. Vertical movements of air depends upon the environmental conditions. If a rising parcel of air arrives at an altitude in a cool or denser state than its surroundings then the displaced air will not move upwards. When there is no movement, harmful pollutants accumulate only at the ground level. Air near the surface of earth cools more than the layers above. This causes less dilution and little movement of air with pollutants in the ambient atmosphere.

Increase in temperature with altitude is known as an *inverse condition* (Figure 7.1), which is unfavourable for the dispersion of pollutants. Valleys, low

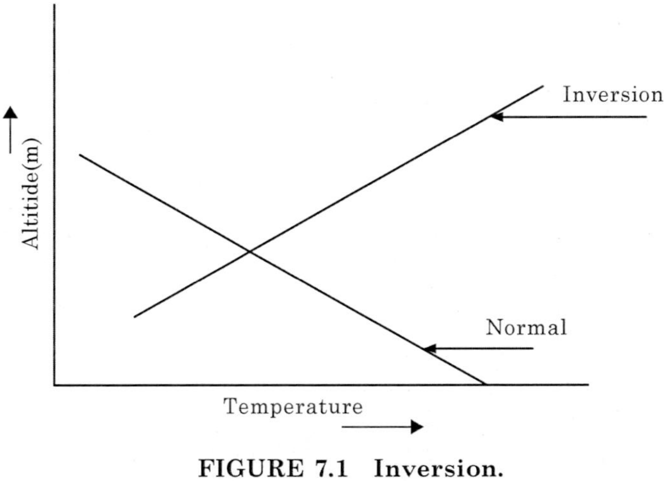

FIGURE 7.1 Inversion.

lying areas and coastal areas are affected as dense and cooler air stays below the warm air. Early morning fog traps gaseous and particulate pollutants. Unequal cooling rates of surface of earth and air layers in contact (before sunrise), air moving down to replace air which has flown out from high pressure regions and hill ranges near sea experience inverse conditions. In Belgium (1930), Los Angeles (1941) and Thames Valley (1962), many people died in a few days due to exposure to the adverse inverse atmospheric conditions.

Affects due to inverse conditions may be reduced if

(i) gaseous pollutants are minimized at source;
(ii) the stack height is adequately increased; and
(iii) care is taken at the time of locating the industries.

Photochemical Smog

Reactive hydrocarbons interact with ozone to form a hydrocarbon free radical which readily reacts with oxygen to form another free radical ($RCH_2 \cdot O_2$), which reacts with NO to produce NO_2 and free radical RCH_2O. When RCH_2O reacts with oxygen, a stable aldehyde RCHO and hydro peroxyl radical HO_2 forms which is extremely reactive and regenerates the hydrocarbon free radical. This goes as a chain reaction. Finally, Peroxy Acetyl Nitrate (PAN) which is a potent and toxic irritant in the smog is formed due to the presence of particulates, NO_2, O_3, HC, volatile organic carbon, CO in the atmosphere. Smoke and fog keep all the harmful secondary pollutants at the ground level.

7.5 ACID RAIN

Industrial activities (combustion or chemical process) release SO_2, NO_2 and other gases into the atmosphere. Fine particulates, aerosols and various salts of acidic origin accumulate in the polluted atmosphere, water vapour condenses on these aerosols and increase in their size. These large size droplets acidic in nature get mixed with rain water. Oxidation takes place in the presence of OH, O_3 or H_2O_2. SO_2 is oxidized to SO_3 and forms sulphuric acid under humid conditions. NO_2 is oxidized to NO_3 or N_2O_5 and forms nitric acid in the presence of moisture. HCl gas may also result into hydrochloric acid. Dry or droplet deposition on plant leaves or building materials in the atmosphere, mixes with precipitation and decreases the pH value of water and soil to 5 or so. Reduction of pH releases harmful chemicals dangerous to human beings.

Adverse Impacts

Adverse effects of acid rain are as follows:

- (i) It causes damage (erosion, corrosion, fading, cracking) to materials like steel, paints, fibrous material, and fabrics.
- (ii) It reduces soil fertility (lack of sufficient nutrients) and crop yield.
- (iii) It affects aquatic organisms, plant life and biodiversity. Some species may be eliminated.
- (iv) It results irritation to skin and respiratory tract.
- (v) It adds toxic chemicals to food products, water bodies and soil.
- (vi) Ancient monuments like Tajmahal gets corroded.

The only available remedy is to prevent or reduce the release of sulphur and nitrogen oxides, into the atmosphere from various industries. Use of cleaner coal (without sulphur) or alternate energy sources may be preferred. Industries may use electrostatic precipitators or catalytic converters to reduce gaseous emissions.

7.6 GREENHOUSE EFFECT

Carbon dioxide concentration in the atmosphere is rapidly increasing every year at the rate of 1 to 2 ppm in developed countries due to industrialization and deforestation. Since water vapour and CO_2 trap (absorb) a large portion of emitted (reflect) radiation (of higher wavelength 4 nm) near ground level, earth gets heated up slowly. Atmospheric temperature is also altered due to the presence of dust, particulate matter and other contaminants. Greenhouse effect results in trapping the heat near ground level (Figure 7.2) increasing global

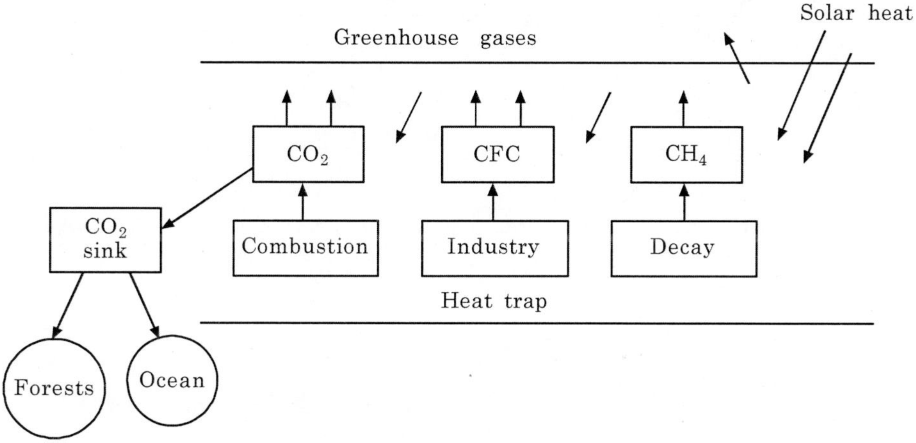

FIGURE 7.2(a) Greenhouse effect.

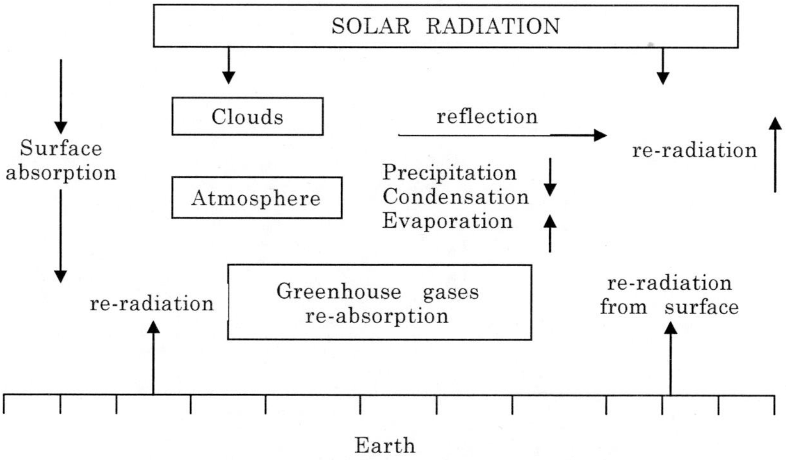

FIGURE 7.2(b) Solar energy radiation.

temperatures by few degrees over a long period of time. Some portion of the infrared radiation near the surface of the earth is needed to maintain a temperature favourable for life on earth. CFCs, CH_4, O_3 and N_2O in the atmosphere also act as a glass chamber which permits the incoming solar radiation of low wavelength but prevents the re-radiated radiations to go out of the enclosure shown in Table 7.1. $CFCl_3$, CF_2Cl_2 released due to human activities, have a residence time over 50 years in the atmosphere and also more responsible for the green house effect. But the quantity of carbon dioxide present in the atmosphere is more to cause a very significant effect on global temperature.

Soil, forests and seas are sinks which remove CO_2 from the atmosphere.

Table 7.1 Significance of different gases

Gas	*Significance* (per cent)	*Source*
CO_2	64	Combustion, deforestation
CH_4	19	Decomposition of organic matter
CFC's	19	Industrial applications
N_2O	6	Fertilizers

Most of the advanced countries like the USA, Europe, and Russia contribute highest percentage of greenhouse gases into the atmosphere. As air moves without boundaries, other countries are also affected.

7.7 OZONE LAYER DEPLETION

Ozone is undesirable at ambient levels because of the following reasons:

(i) It affects eyes and respiratory tract.
(ii) It results in green house effect.
(iii) It causes photochemical reactions, the end products being more harmful.

At ground levels, ozone will be in traces only. In stratosphere, Ozone layer shield is desirable as it prevents solar (ultraviolet) radiation into the atmosphere, which is harmful to plants and animals. Figure 7.3 shows ozone layer.

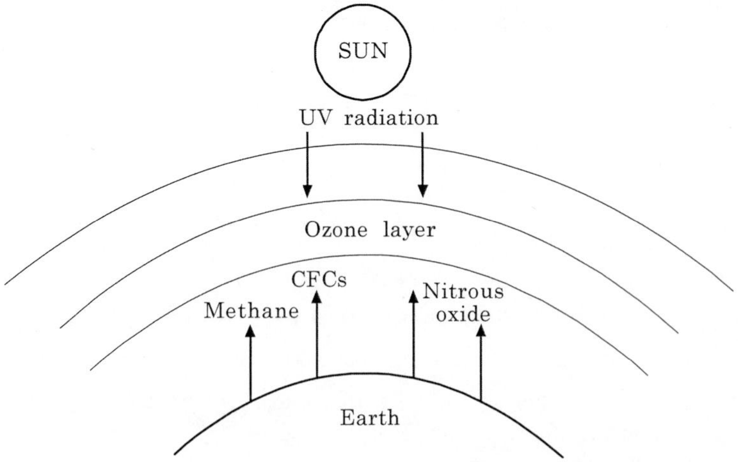

FIGURE 7.3 Ozone layer.

Oxygen molecule dissociates into two oxygen atoms under solar ultraviolet radiation. Then oxygen atoms combine with oxygen molecules present in the atmosphere to form ozone. Atomic oxygen is also released by ozone. The process of ozone formation is a stable and continuous phenomena.

$$O_2 + O \rightleftarrows O_3$$

NO, Cl and OH also react with ozone.

$$\left.\begin{aligned} NO + O_3 &\rightarrow NO_2 + O_2 \\ NO_2 + O &\rightarrow NO + O_2 \end{aligned}\right\}$$

$$\left.\begin{aligned} Cl + O_3 &\rightarrow ClO + O_2 \\ ClO + O &\rightarrow Cl + O_2 \end{aligned}\right\}$$

$$\left.\begin{aligned} OH + O_3 &\rightarrow HO_2 + O_2 \\ HO_2 + O &\rightarrow OH + O_2 \end{aligned}\right\}$$

In all these reactions which are very complex in nature, NO, Cl, and OH continue to exist in the atmosphere and reduce the thickness of the ozone layer. Thickness of this useful ozone layer at different latitudes and at different positions in the stratosphere is likely to vary from 5 to 10 ppm. It is maximum near polar region and the destruction is also maximum.

Causes of Depletion

Human activities in developing countries release $CFCl_3$, CF_2Cl_2 (refrigeration and air conditioning units) HCFCs, HBFCs (fire extinguishers), CCl_4, methyl chloroform, nitrogen oxides, CO_2 etc. into the atmosphere. Chlorine atoms are released into the atmosphere from CFCs (chlorofluoro carbons). These chlorine atoms react and destroy the ozone molecules. Chlorine and other reactants have a residence time of 50 to 100 years in the atmosphere and are quite stable in nature.

$$\begin{aligned} CFCl_3 + \text{UV energy} &\rightarrow CFCl_2 + Cl \\ Cl + O_3 &\rightarrow ClO + O_2 \\ O_2 + \text{UV energy} &\rightarrow 2O \\ ClO + O &\rightarrow O_2 + Cl \end{aligned}$$

Thus chlorine, the main reactant is always available to deplete the ozone layer concentration. When the thickness is reduced, it leads to the formation of a 'hole'; which allows the rays of Sun to penetrate directly into the biosphere.

Harmful Effects

Absence of protective ozone layer causes the following harmful effects:

(i) Mutation and genetic abnormalities, skin cancer, eye and other infectious diseases

(ii) Harm the ecosystems, plants, animals and human beings

(iii) Increase in temperature on the earth
(iv) More exposure to solar radiation
(v) Formation of photochemical smog due to secondary reactions in the atmosphere because of presence of O_3 in ambient atmosphere.

Remedies

Most of the chemicals reacting with ozone are released only from refrigeration and air conditioning units, foams, thermoware, paints, pesticides, perfumes and pharmaceutical industry. Hence alternative technologies should be used so that chlorine and other harmful substances are not released from those industries. Limitations were put forward on the use of CFC's and halons in the Montreal protocol (1987) and consumption reduced as reported by UNEP (1992). HFC, HCFC may be preferred to CFC's in refrigeration, air-conditioning and foam products as they have less residence time in the atmosphere. One atom of chlorine destroys 1 lakh molecules of ozone.

7.8 GLOBAL WARMING

Climate is not uniform at all places on the earth. But is almost the same at a particular location. Forests, deserts, floods, droughts, mountains, etc. are all natural features in the global environment. But bio-geo-chemical cycles in nature balance the entire ecosystem. Almost the temperature on earth appears to be constant. However, observations reveal an increase of 0.5°C temperature and 5 to 10 cm rise in sea level over a period of 100 years. This increase appears to be at a slow rate but is likely to produce alarming conditions. CO_2, CH_4, CFC's, N_2O and other chemicals, dust, particulate matter released from the industries in the developed or developing countries, absorb infrared radiation and warm the atmosphere at ground level. Greenhouse gases, ozone depletion and pollution levels add to the cause of global warming.

Contribution to global warming due to deforestation, industrial units, and excess energy use are also due to human activities.

Global temperatures will not increase uniformly at all places. Wind movements, rainfall and climatic changes also have different patterns. These differences affect plants and animal life at all places. Further, when the atmospheric temperatures increases water vapour content in the atmosphere is going to be more due to evaporation and transpiration. Presence of more water vapour again causes higher temperatures.

Harmful Effects

The harmful effects of global warming are as follows:

(i) Climatic changes
(ii) Melting of ice
(iii) Rise in sea level

(iv) Submergence of coastal wetlands
(v) Changes in crop patterns
(vi) Dispersion of harmful chemicals
(vii) Effects on ecosystems and biodiversity
(viii) Changes in hydrological cycle
(ix) Changes in soil characteristics

All these changes are likely to be compensated by a process known as *global dimming*. Scientists have observed 2 to 4 per cent reduction in the amount of solar radiation reaching the earth's surface, with the increase in cloud cover, aerosols and particulates in the atmosphere. Higher temperatures lead to an increased cloud cover. The scattered light through the clouds boosts the plant's absorption of carbon dioxide and photosynthesis process. Thus global dimming is a process working against global warming to some extent.

7.9 HAZARDOUS WASTES

Toxic wastes cause significant harm to human health or ecosystems when improperly stored, transported, treated or disposed. Wastes from radioactive and explosive units are extremely hazardous as they cause irreversible damage to the environment. Cyanides, mercury, arsenic, lead, cadmium, chromium, selenium even in very small concentrations (< 0.05 ppm) are very harmful to the living beings. Wastes from paints, dyes and other processing units, oil and emulsions, sludges carrying heavy metals, acid or alkali slurries are very harmful. These wastes are highly volatile (ignitable), corrosive (extreme pH), reactive (explosive) and highly toxic in nature.

It is always preferable to use alternate products, materials, processing technologies to minimize the generation of toxic wastes. They may be detoxified or converted into a stable state. Some options available are, evaporation, centrifugation, absorption, adsorption, reverse osmosis, filtration, precipitation, neutralization, oxidation, reduction, ion exchange, solidification, stabilization, fixation, biodegradation, electrodialysis and thermal destruction. Hazardous wastes are disposed on land or by incineration after final treatment.

Wastes are solidified, immobilized, compacted, dewatered and stored in controlled land fills. Impermeable clay liners prevent the leachate flow and soil contamination. (Figure 7.4)

Thermal oxidation (Figure 7.5) is carried at very high temperatures (1000°C) for the disposal of organic wastes, pesticides, oils, and biologically hazardous wastes.

Nuclear Wastes

Weathering of atomic minerals (Thorium, Uranium, Radium) from the earth's crust is a natural source for radioactive substances. These minerals are found to be more useful in power sector, industry and medical field. They reach air, water and soil and enter the food chain causing serious damage to all living things.

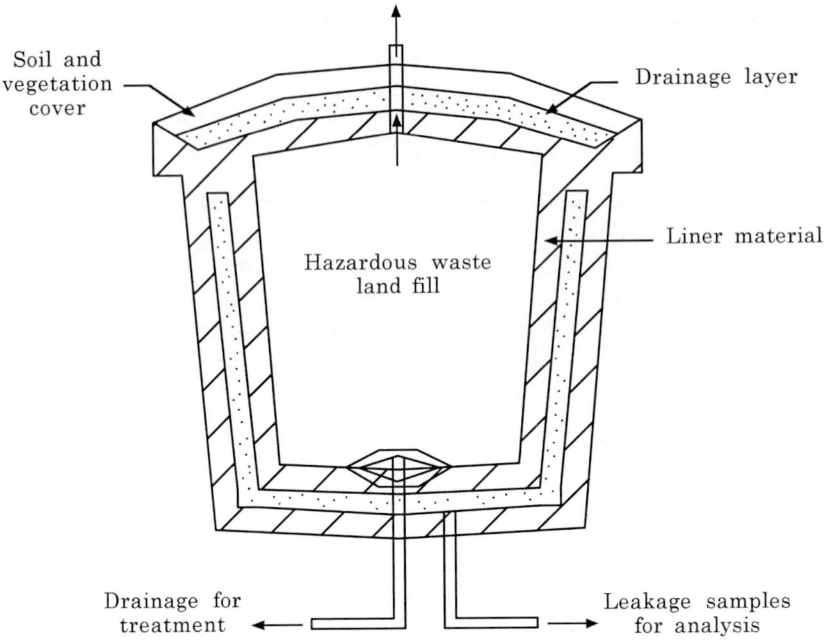

FIGURE 7.4 Land fill.

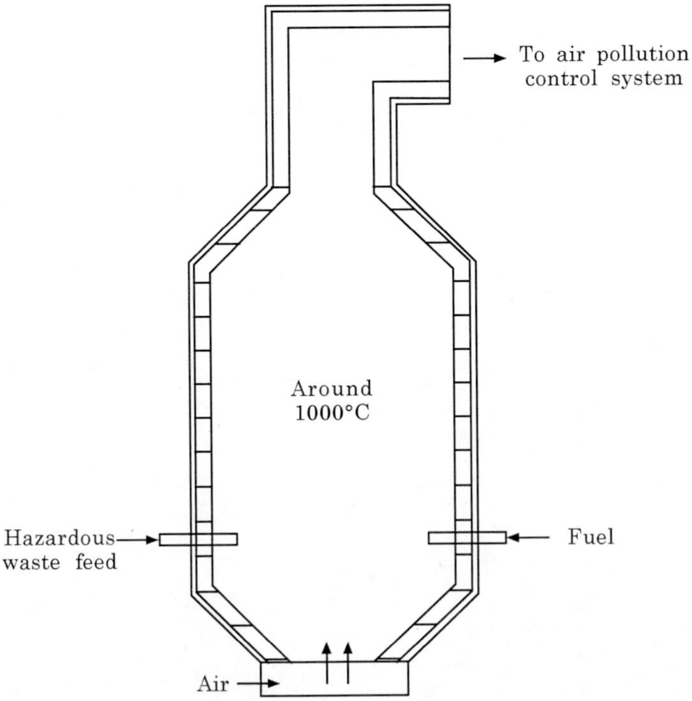

FIGURE 7.5 Incinerator.

They cause many physiological changes in the subsequent generations as evident from Hiroshima episode and Chernobyl explosion. Depending upon the dosage and exposure duration, short term or long term, temporary or permanent effects may happen due to radiation levels in the environment. Contaminants may be released into the atmosphere due to mining activities or leakage due to system faults in the reactors.

A number of methods are available for the disposal of radioactive wastes.

(i) Solidification and immobilization or evaporation and concentration to contain the waste so that they may be buried ultimately in underground special alloy containers or in deep sea waters.
(ii) Very low concentration radioactive substances may be diluted further and disposed in marine waters.
(iii) Wastes may also be imprisoned in concrete lined underground tanks for decay over a long period of time.

7.10 DISASTER MANAGEMENT

Earthquakes, landslides, cyclones, floods, explosions, epidemics, nuclear accidents may be identified as disasters causing instantaneous damage on a large scale in the environment. They cannot be sometimes predicted but areas prone to each type of disaster can be identified on the basis of earlier records. A devastation causes the following:

(i) Human, animal and plant life or the entire ecosystem is lost.
(ii) Buildings and infrastructural facilities are severely damaged.
(iii) Road, railway and other communication systems collapse completely.
(iv) Water supply and power systems fail.
(v) Crops and cattle feed get spoilt beyond use.
(vi) Cholera, plague and other epidemics spread.
(vii) Economic and social disturbance increases.

Rehabilitation is a very difficult task and requires a lot of money, effort and time to restore normalcy.

Preventive Measures

(i) Landslides can be reduced along the mountain slopes by providing suitable drainage measures and preventing deforestation.
(ii) Floods can be prevented by constructing suitable bunds and developing vegetation along the banks.
(iii) Special design procedures and construction practices can take care of impacts from cyclones or earthquakes.

Community shelters provide safe accommodation during disasters. Things to be ready in the event of disaster are as follows:

(i) Home guards for distribution of food and medicines

(ii) Fire, medical and ambulance services
(iii) Information centres to create public awareness

Measures needed are as follows:

(i) Rescue and relief operations
(ii) Restoration of power, water and communication systems
(iii) Supply of food, medicine and other basic needs
(iv) Maintenance of sanitation
(v) Safeguarding agricultural products, livestock and public properties

All volunteers should act as a single team utilizing the available resources to help the victims.

SUMMARY

Population densities are high in cities and urban slums. The associated problems are traffic congestion, solid waste disposal, less vegetation, water supply and drainage problems and higher costs of living.

Solid, liquid and gaseous wastes are released into the atmosphere from various industries polluting natural resources. Gaseous pollutants accumulate in the ambient atmosphere if air movements are restricted due to inversion. Ground temperature increases if carbon dioxide and other green house gases are more in the atmosphere or ozone layer depletes in the stratosphere. Wastes containing cyanides, arsenic and other toxic chemicals are more hazardous in nature. They should be concentrated and contained properly deep in the ground. Thermal oxidation at very high temperatures may be very effective for safe disposal of some highly toxic pollutants.

Areas prone to different disasters like earthquakes, floods, need to be identified for taking suitable preventive or rehabilitation measures.

KEYWORDS

Acid rain, agricultural revolution, disaster mitigation, global warming, green house effect, hazardous nuclear wastes, industrialization, inversion, ozone layer depletion, urbanization.

SOLVED QUESTIONS

1. **List out the specific problems due to unchecked urban growth?**

   Some are due to overcrowding, reduction in open space, traffic congestion, cost of living, increasing rental values, transportation requirements, inadequate water supply, noise disturbance, sharing of infrastructural facilities. All these lead to stress on population.

2. Advanced countries are contributing more of CO_2, CFC's, SO_2, NO_2, CH_4. Any comments?

Man is altering the environment for his comforts, forgetting the fact that life supporting systems have emerged through a gradual process of evolution and are in a state of delicate balance. Air conditioning, refrigeration, foam preparations, fire extinguishers, solvents, aerosols, electrical and metal processing units release CFC's and Halons, which are stable and active for even 100 years in the atmosphere. Fuel burning and deforestation are the two main reasons for more CO_2 in the atmosphere. Supersonic aircrafts release NO_x. 50 per cent of these harmful emissions are coming out from USA, Canada, Russia and European countries, since last 50 years. Due to ozone depletion or global warming, it is likely that underdeveloped or developing countries are to suffer more than the affluent advanced countries. The following steps are being taken to reduce the harmful emissions into the atmosphere as per the international agreements:

(i) Continuous monitoring of the atmosphere
(ii) Limitations on the use of harmful chemicals
(iii) Phasing out these chemicals
(iv) Use of alternate technologies and relatively safer chemicals like HFC, HCFC instead of CFC, Halons.

3. How are radioactive wastes disposed?

Wastes are concentrated and contained in a controlled operations of burial (land or sea) or burning at very high temperatures.

4. Mention the human activities that are likely to create disasters?

(i) Agricultural operations and construction activities on the hill slopes
(ii) Removal of vegetation along the banks of water bodies
(iii) Construction of large dams and reservoirs
(iv) More withdrawal of groundwater

REVIEW QUESTIONS

1. What are the steps to curb urbanization trends?
2. Explain how green house gases increase global temperatures?
3. What are the disadvantages due to depletion of ozone in the stratosphere?
4. Mention the characteristics of hazardous wastes.
5. Discuss the merits and demerits of nuclear power plants.
6. Write a note on the first steps to be taken after a disaster.
7. What do you understand by 'solar radiation'?

OBJECTIVE TYPE QUESTIONS

Fill in the blanks with suitable words:

1. Inversion is unfavourable for the __________ of pollutants in the atmosphere.
2. Acid rain is due to ______________.

   (a) CO_2 (b) SO_2 (c) CFC (d) NH_3
3. Depletion of ozone is due to

   (a) Chlorine (b) Carbon (c) HFC (d) none
4. Floods can be prevented by constructing ______________.
5. The main contributers in global emission of CO_2 are _________ countries.
6. Thickness of ozone layer is measured in ___________ units.
7. When a layer of warm air settles over a layer of cooler air, the condition leads to

   (a) inversion (b) global warming (c) acid rain (d) dispersion
8. Ozone layer absorbs__________________.

   (a) UV radiation (b) infrared radiation (c) radio waves (d) chlorine atoms
9. Ozone concentration is ______________ in stratosphere.

   (a) 10 ppm (b) infrared radiation (c) constant (d) nil

CHAPTER 8

ENVIRONMENTAL MANAGEMENT

TOPICS COVERED

- Environmental Impact Assessment
- Waste as a Resource
- Environmental Laws
- Environmental Management Plan
- Policies for Quality Improvement

8.1 ENVIRONMENTAL IMPACT ASSESSMENT

Projects like construction of a dam on a river, nuclear power plant, fertilizer plant are always taken up for the benefit of the society. But these developmental activities may also produce adverse impacts on the environmental systems, if proper remedial measures are not adopted. Depending upon the height of the dam, reservoirs may submerge a number of villages. An industry may pollute the surrounding areas unless wastes are properly treated before disposal into water bodies or on the land. The main purposes of environmental impact assessment study are as follows:

(i) To analyze the adverse effects due to all project activities, on the total environment, in a systematic manner

(ii) To modify the project activities or to implement waste management programmes to minimize pollution levels before disposal

(iii) To select the best alternatives which have minimum adverse effects on the environment

Major project proposals will not be approved by the government, unless they satisfy the environmental considerations.

A detailed study is required based upon the following aspects:

(i) Status of the environment
(ii) Project activities
(iii) Environmental parameters affected
(iv) Pollution abatement measures
(v) Overall evaluation of the damage function

The existing environmental conditions with regard to natural resources like water, land, air, forests, minerals etc., socio-economic living standards, and presence of historic or cultural moments, can always be correctly documented. The present status of the environment and quality of life should not be destroyed directly or indirectly after the implementation of a new project. All the project details including construction activities, procurement of raw materials, manufacturing processes, waste generation and disposal systems, should be understood so as to estimate the quantum of damage or depletion of natural resources and environmental systems. Feasibility of alternate technologies in the process of manufacturing units or change in raw materials or providing control devices to prevent or minimize pollution levels must be thoroughly examined.

Finally, number of adverse effects on the environment due to a number of project activities are to be analyzed, evaluated and quantified in detail. Typical project environment interaction from an industry is shown in Figure 8.1(a).

Methodology

Effects on the environment due to different project activities that can be presented in a tabular column are shown in Figure 8.1(b). It is advisable to

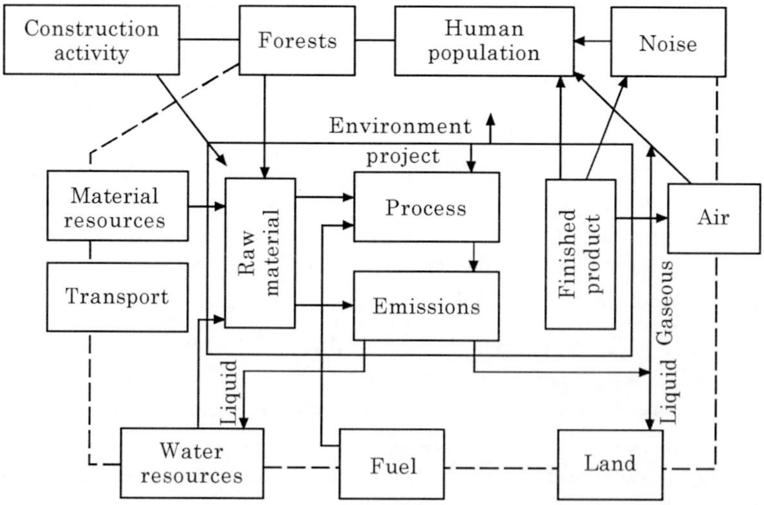

FIGURE 8.1(a) Project-environment interaction.

ENVIRONMENTAL ATTRIBUTES

Project activities	*Water*		*Land*		*Air*		*Social*		*Others*	
	Quantity	Quality	Land use patterns	Soil characteristics	Particulates	Gaseous emissions	Employment	Cultural	Noise	Aesthetics
Site conditions										
Construction activity										
Raw materials										
Fuel needs										
Processing units										
Reaction towers										
Sources of waste generation										
Pollution control devices										
Disposal of emissions										

FIGURE 8.1(b) Project-environment interaction.

divide and subdivide various project activities further and further to the extent possible so that the alternate evaluation on the total harmful effect on the environment is more accurate. Similarly, environmental attributes of concern are also subdivided further to know how environmental quality is affected in detail,

Parameters may be listed as:

Physical environment (topography, climate, geological, hydrological, soil characteristics)
Land use (agriculture, forestry, fishing)
Infrastructure (human settlements, communications)
Social factors (demography, health care, education facilities, employment opportunities, income levels)
Water sources (depletion, quality, toxic substances)
Air pollution (CO, SO_2, NO_2, HC, particulate matter)
Ecosystems (crop patterns, vegetation, aquatic life, species diversity)
Others (noise, odour, historic or cultural value)

Project activities may be listed as:

Raw materials (quantity, composition, source)
Process (production flow charts, main reactions, byproducts)
Water, fuel and energy requirements
Solid wastes generated (quantity and characteristics)
Liquid effluents (sources, quantities, characteristics)
Gaseous emissions (sources, nature and composition)
Management plans to mitigate pollution levels
Noise levels, odour, thermal pollution
Occupational hazards, sanitation, safety measures

In this matrix form of chart, it is important to identify whether a particular project activity is creating any harmful effect on any aspect of environment. The extent of damage has to be identified. The damage function may be quantified in different ways and is marked in the tabular columns corresponding to the project activity and environmental attribute. The damage may be a short term or long term effect. It may be irreversible, local, insignificant or secondary in nature. Various impacts are categorized and assigned as units of importance. It can be on 0 to 1, or 1 to 10 scale (good to bad). These parameters are likely to change depending upon the alternate technologies for processing the chemicals or introduction of pollution control devices as required. All these units of quantification summed up as an environmental index presents clearly the environmental impact statement.

Notes:

(i) All parameters are not of equal significance.
(ii) Even if few effects are very harmful, the project should be discontinued.
(iii) Public opinion is needed in E.I.A. study.
(iv) It is difficult to combine the different items into one simple index.

Another method of E.I.A. is network system analysis of studying the nature and effect of various related parameters as shown in Figure 8.2(a) for reservoir development and in Figure 8.2(b) for power plant project.

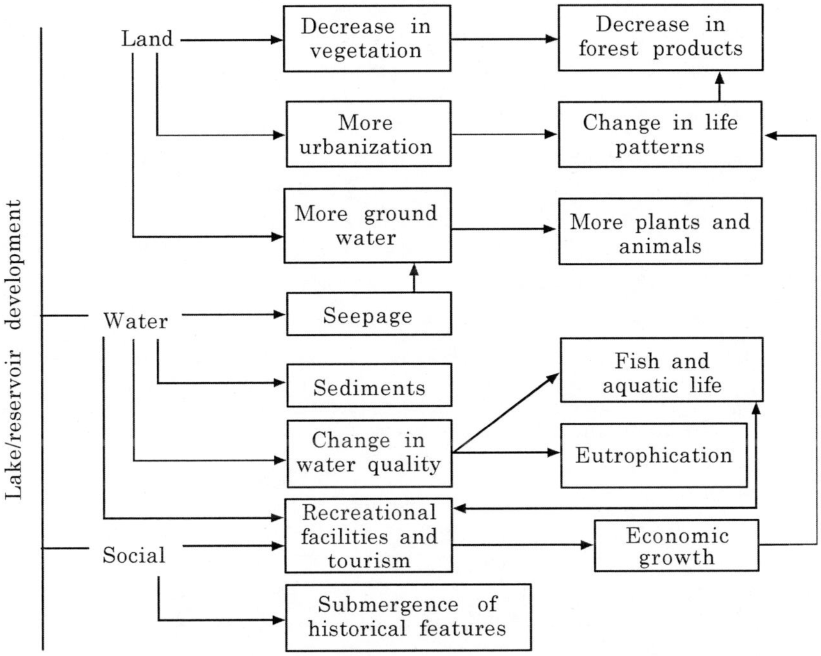

FIGURE 8.2(a) Network system.

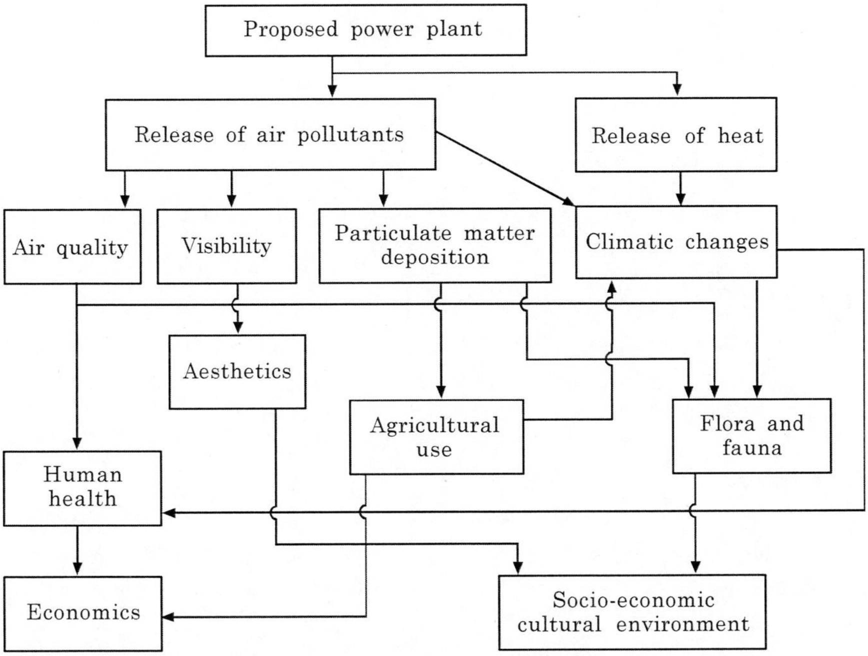

FIGURE 8.2(b) Network analysis on air environment.

8.2 WASTE AS A RESOURCE

Waste is recognized as a resource which helps in the sustainable development of the environment. Some of the waste materials can be recycled or recovered and used again in the manufacturing process of the same product or a new commercial product. Some examples are specified below.

(i) Useful materials from waste products:

- Bricks using flyash, paper boards from rice husk, bagasse
- Wood substitute panels from cotton stalks
- Preparation of concrete using flyash, replacing 10 per cent cement
- Use of refinery sludge as soil conditioner
- Inert solid wastes as a fill material for slush lands

(ii) Re-use of materials from process units:

- Heat value from incineration of solid wastes
- Biogas from digesion plants
- Cooling waters
- Fluorides from aluminium industry
- Pulp from paper manufacturing process
- Remoulding plastic material
- Recovery of ammonia
- Gaseous products from petroleum refineries

(iii) Financial gains from left overs:

- Proteins from slaughter house
- Manure from composting units
- Briquettes from agricultural residues
- Fuel pellets from garbage
- Animal feed from food industry
- Vegetable oils from sugar molasses

(iv) Recovery of materials from processing units:

- Silver from photographic filming
- Zinc from rayon plant
- Iron from steel works
- Solvents from chemical units
- Cu, Pb from cables and other units
- Sulphur from power plants
- Hydrocarbons from petrochemical residues

All the aforementioned examples are taken directly from the industrial units, which are successfully using waste as a resource. These procedures minimize waste production and are cost effective. That is the reason for introducing these practices in the environmental management plans.

8.3 ENVIRONMENTAL LAWS

Requirements of a Contract

Contract is an agreement to perform some obligation with due consideration to the conditions which can be enforced by law.

(i) It must be between two persons or parties.
(ii) The nature of work must be a feasible project.
(iii) There must be complete understanding on the job to be undertaken.
(iv) The work must be completed within a reasonable time.
(v) Equitable consideration is essential after job completion.
(vi) Valid documentation is needed.
(vii) It cannot be between minors or mentally unsound persons.
(viii) The subject cannot be a social obligation or intentions to do wrong things.
(ix) Innocence and ignorance on some issues may not be excused.

All procedures and code of practices, rights and duties form a part of the law. All legal provisions are enacted in the relevant acts by the government. When one party has performed the duty, non-compliance by the other is punishable with fine or imprisonment or both. State and Central Governments have the power to make rules and amendments from time to time. Courts (under proper jurisdiction) will take the cognizance of offence and issue suitable directions to implement the legal provisions of an act. The contract is void or voidable if there are basic flaws in fulfilling the conditions of contract. For causes of general interest, public may also approach the court for justice. Finally the good of the people is the supreme law.

Environmental Legislations

It shall be the duty of every citizen of India to protect and improve the natural environment including forests, lakes, rivers, wildlife and have compassion for all living creatures. Right to life is guaranteed under Indian Constitution.

Stockholm Conference (1972) has outlined the following **Principles** with regard to environmental protection.

(i) Protect the environment and natural resources as quality living and social well being is the fundamental right of the present and future generations. It includes conservation of wildlife, marine life, other ecosystems, nonrenewable and natural resources. The carrying capacity of the earth is limited.
(ii) Create better living conditions (socio-economic development) through technological advances and management practices.
(iii) Keep proper control over population growth and density of population.

(iv) Stop the discharge of toxic substances, which causes irreversible damage to the ecosystem or create adverse effect on the quality of environment.

(v) Encourage scientific-research activities and create environmental awareness.

(vi) Make and use laws to take care of damage to environment and victims from pollution hazards.

It is the duty of the government to regulate and monitor environmental quality and punish the persons who endanger the human environment knowingly or unknowingly, directly or indirectly with pollution hazards, loss of forests and biodiversity by releasing harmful chemicals into the atmosphere effecting the life supporting systems. Hence to protect and improve the natural environment and right to life (human beings, plants, microorganisms, property, wild life, forests, lakes, rivers etc.), Water Act, 1974, Air Act, 1981 and Environmental Protection Act, 1986 have been promulgated by the government. In addition, Forest Conservation (1980), Wild Life Protection (1972), Factories (1948, 1987), Hazardous Wastes (1989), Motor Vehicles (1989), Public Liability Insurance Act (1991) and other procedures under Indian Penal Code are applicable for preventing pollution, maintaining ecological balance, conserving natural resources and safeguarding the interests of all living organisms.

The Acts contain the following provisions:

- Definitions of different technical terms such as pollutant, approved appliances, emission standards, etc.
- Rules prohibiting the discharge of domestic and industrial effluents with harmful chemicals
- Penal provisions in case of indiscriminate discharge of wastes with inadequate treatment

Water Act is connected with the discharge of liquid/solid effluents into water sources affecting water quality, aquatic life, land irrigation whereas Air Act is related to the cause and effects of air pollution. Handling, storing, transporting of hazardous chemicals and the safety of workmen are also included in the relevant acts.

To implement the legal provisions of the act and to deal with environment related issues of the act, Pollution Control Boards are established from government funds at State and Centre levels with a Chairman, Secretary, and Members to represent agriculture, fisheries, industries, trade and other areas of interest. They may constitute sub-committees for any specific purpose. Members from local bodies may also be co-opted. Joint Boards may be set up for solving bilateral issues. Central Pollution Control Board will have the jurisdiction over the entire nation and co-ordinates all matters with regard to various State Boards.

Government also classified the industries into different (red, orange and green) categories depending upon the presence of harmful chemicals in the raw materials used or the manufacturing processes so that special care can be taken with regard to those industries which are more dangerous. Pollution Control Board gives consent to industries only after satisfying with the location and periodically inspect them for violation of pollution control measures.

THE ENVIRONMENTAL (PROTECTION) ACT, 1986 No. 29 of 1986

"EXTRACTS"

CHAPTER I

PRELIMINARY

1. SHORT TITLE, EXTENT AND COMMENCEMENT

(i) It extends to the whole of India.

2. DEFINITIONS

In this act, unless the context otherwise requires,

(a) **"Environment"** includes water, air and land and the inter-relationship which exists among and between water, air and land, and human beings, other living creatures, plants, micro-organism and property.

(b) **"Environmental pollutants"** means any solid, liquid or gaseous substance present in such concentration as may be, or tend to be, injurious to environment.

(c) **"Environment pollution"** means the presence in the environment of any environmental pollutants in the environment.

(d) **"Handling",** in relation to any substance, means the manufacture, proceedings, treatment, package, storage, transportation, use, collection, destruction, conversion, offering for sale, transfer or the like of such substance.

(e) **"Hazardous substance"** means any substance or preparation, which by reason of its chemical or physio-chemical properties or handling, is liable to cause harm to human beings, other living creatures, plant, micro-organism, property or the environment.

(f) **"Occupier",** in relation to any factory or premises, means a person who has, control over the affairs of the factory or the premises and includes in relation to any substance, the person in possession of the substance.

(g) **"Prescribed"** means prescribed by rules made under this act.

——— * * * ———

CHAPTER II

GENERAL POWERS OF THE CENTRAL GOVERNMENT

POWER OF CENTRAL GOVERNMENT TO TAKE MEASURES TO PROTECT AND IMPROVE ENVIRONMENT

1. Subject to the provisions of this act, the Central Government, shall have the power to take all such measures as it deems necessary for the purpose of protecting and improving the quality of the environment and preventing controlling and abating environmental pollution.
2. In particular, and without prejudice to the generality of the provisions of sub-section (1), such measures may include measures with respect to all or any of the following matters, namely:

   (a) Co-ordination of actions by the State Governments, officers and other authorities:

      (i) under the act, or the rules made thereunder or
      (ii) under any other law for the time being in force which is relatable to the objects of this act.

   (b) Planning and execution of a nation-wide programme for the prevention, control and abatement of environmental pollution.
   (c) Laying down standards for the quality of environment in its various aspects.
   (d) Laying down standards for emission or discharge of environmental pollutants from various sources whatsoever.
   (e) Restriction of areas in which any industries, operations or processes or class of industries, operations or processes shall not be carried out or shall be carried out subject to certain safeguards.
   (f) Laying down procedures and safeguards for the prevention of accidents which may cause environmental pollution and remedial measures for such accidents.
   (g) Laying down procedures and safeguards for the handling of hazardous substances.
   (h) Examination of such manufacturing processes, materials and substances as are likely to cause environmental pollution.
   (i) Carrying out and sponsoring investigations and research relating to the problems of environmental pollution.
   (j) Inspection of any premises, equipment, machinery, manufacturing or other processes, materials or substances and giving order of such directions to such authorities, officers or persons as it may consider necessary to take steps for the prevention, control and abatement of environmental pollution.
   (k) Establishment or recognition of environmental laboratories and institutes to carry out the functions entrusted to such environmental laboratories and institutes under this act.

(l) Collection and dissemination of information in respect of matters relating to environmental pollution.
(m) Preparation of manuals, codes or guides relating to the prevention, control and abatement of environmental pollution.
(n) Such other matters as the Central Government deems necessary or expedient for the purpose of securing the effective implementation of the provisions of this act.

RULES TO REGULATE ENVIRONMETAL POLLUTION

The Central Government may, by notification in the official gazette, make rules in respect of all or any of the matters referred to

In particular, and without prejudice to the generality of the foregoing power, such rules may provide for all or any of the following matters, namely:

(a) The standard of quality of air, water or soil for various areas and purposes
(b) The maximum allowable limits of concentration of various environmental pollutants (including noise) for different areas
(c) The procedures and safeguards for the handling of hazardous substances
(d) The prohibition and restrictions on the handling of hazardous substances in different areas
(e) The prohibition and restriction on the location of industries and the carrying on process and operations in different areas
(f) The procedures and safeguards for the prevention of accidents which may cause environmental pollution and for providing remedial measures for such accidents

PREVENTION, CONTROL, AND ABATEMENT OF ENVIRONMENTAL POLLUTION

No person carrying on any industry, operation or process shall discharge or emit or permit to be discharged or emitted any environmental pollutants in excess of such standards as may be prescribed.

No person shall handle or cause to be handled any hazardous substance except in accordance with such procedure and after complying with such safeguards as may be prescribed.

POWERS OF ENTRY AND INSPECTION

Subject to the provisions of this section, any person empowered by the Central Government in this behalf shall have a right to enter, at all reasonable times with such assistance as he considers necessary, at any place.

For the purpose of examining and testing any equipment, industrial plant, record, register, document or any other material object for conducting a search of any building in which he has reason to believe that an offence under this act or the rules made thereunder has been or is being or is about to be committed and for seizing any such equipment, industrial plant, record, register, document or other material object if he has reason to believe that it may furnish evidence of the commission of an offence punishable under this act or the rules made thereunder or that such seizure is necessary to prevent or mitigate environmental pollution.

POWER TO TAKE SAMPLE AND PROCDURE TO BE FOLLOWED IN CONNECTION THEREWITH

The Central Government or any officer empowered by it in this behalf, shall have power to take, for the purpose of analysis, samples of air, water, soil or other substance from any factory premises or other places in such a manner as may be prescribed.

ENVIRONMENTAL LABORATORIES

The Central Government may, by notification in the official Gazette:

(i) Establish one or more environmental laboratories.
(ii) Recognize one or more laboratories or institutes as environmental laboratories to carry out the functions entrusted to an environmental laboratory under this act.

PENALTY FOR CONTRAVENTION OF THE PROVISIONS OF THE ACT AND THE RULES, ORDERS AND DIRECTIONS

Whoever fails to comply with or contravenes any of the provisions of this act, or the rules made or orders or directions issued thereunder, shall, in respect of each such failure or contravention, be punishable with imprisonment for a term which may extend to five years with fine which may extend to one lakh rupees, or with both, and in case the failure or contravention continues, with additional fine which may extend to five thousand rupees for every day during which such failure or contravention continues after the conviction for the first such failure or contravention.

OFFENCES BY COMPANIES

Where any offence under this act has been committed by a company, every person who, at the time the offence was committed, was directly in charge of, and was responsible to, the company for the conduct of the business of the company, as well as the company, shall be deemed to be guilty of the offence and shall be liable to be proceeded against and punished accordingly.

* * *

Powers and Functions of Pollution Control Boards

The following are some of the major responsibilities of a PCB:

(i) Advise the government on any matter concerned with prevention, control or abatement of water, air, land or any type of pollution.
(ii) Declare a particular area of land or water body as a no pollution zone and indicate suitability of location site for an industry.
(iii) Collect and disseminate technical information with regard to any or all aspects of pollution.
(iv) Prescribe the standards of water and air quality from time to time with regard to environmental pollutants.
(v) Establish or recognize a laboratory to test samples of water/air.
(vi) Enter and inspect (equipment, process, or documents) any industry so as to assess the levels of pollution from various sources after proper sampling procedures.
(vii) Ensure standards of emission from automobiles.
(viii) Encourage research, training and awareness programmes.
(ix) Provide technical guidance and assistance towards environmental management.

- Pollution Control Board is authorized to issue directions (i) to regulate, prohibit or close any industrial process or operation and (ii) to regulate or disconnect water, electricity or any other service; when the emission of pollutants are in excess of the prescribed standard limits.
- Failure to comply with the directions of PCB in connection with pollution is punishable with fine or imprisonment or both.

These Boards function under the Ministry of Environment which is finally responsible for all the policy framing and implementation. They issue clearance for projects or industries (new, modification or expansion) and co-ordinate with all the concerned departments.

Case Studies

1. M/s X wanted to establish their industries within the catchment basin of Himayatsagar and Osman Sagar lakes. As per the directions of the Supreme Court of India, Pollution Control Board identified and listed red and orange category of industries within the 10 km radius of the aforementioned lakes. The State Government also constituted a high power committee to look into the matters of relocation or closure of industries identified as polluting. Action was taken against those industries accordingly.
2. M/s X established a mechanized slaughter house, which is resulting in the depletion of cattle population. The Government of Andhra Pradesh filed the census of cattle population figures to the Supreme Court of India. The case is pending final disposal.

3. Andhra Pradesh High Court directed the district authorities to remove the illegal encroachments in the surroundings of a lake.
4. Commissioner of 'X' Municipality has been directed to consider the grievances of the locality regarding breakdown of civic amenities and maintenance of public health.
5. Pollution Control Board has issued closure order to the lead extraction unit of M/s X as effluents are released into air/land. Another chemical industry is pooling of effluents posing a grave danger to the local people and nearby water source. Court directed APPCB to act as per law.
6. Court directed the APPCB to strictly enforce E.P. Law with regard to aquaculture projects along the sea shore.
7. Court directed the APPCB not to issue clearance to XX industry till preventive measures of pollution are adopted.
8. Court directed the authorities to see that no garbage (including plastic scrap) is burnt in any thickly populated residential area without obtaining clearance from APPCB.
9. Supreme Court says no to mining in a bid to preserve the ecology of the sensitive Aravalli Hills, spread over Delhi, Haryana and Rajasthan and constituted a high level monitoring committee to suggest ways and means for its overall ecological restoration and suggest remedial measures.

Cases with regard to Bhopal gas tragedy, Almatti dam height, Taj corridor etc. are well known.

8.4 ENVIRONMENTAL MANAGEMENT PLAN

Environmental Audit

E.I.A. is a pre-requisite before setting up an industry. Organization should take care of their own project or industry when once it becomes operative. The management has to systematically and continuously aim to regulate the process within, to safeguard the outside environment. The procedure to review internally the management practices for the evaluation and elimination of pollution liability is known as *environmental audit*. It is an objective, periodic, self evaluation document to know how the processes and equipment are performing to minimize the effluent strength. It is a job to assess the hidden risks, potential hazards and liabilities to avoid accident, leakage, loss and to effectively prevent and control pollution. Internal auditing system helps in safeguarding the health and safety of industrial workers. A manager can implement plans to recover or recycle useful materials.

Audit procedure needs a study of the following:

(i) Raw materials used, details of process and unit operations, products and byproducts produced, quality and quantity aspects, water and material balance and input and output flow charts
(ii) Pollution levels in the surrounding areas in various units
(iii) Evaluation of pollution control systems including waste minimization, recovery and reuse of materials
(iv) Effluent qualities with reference to emission standards
(v) Action plan to improve the efficiency of the whole system

Environmental Management Plan

The basic facts of sustainable development, which have to be understood clearly, are the following:

(i) The basis of human existence is interdependency and maintaining biodiversity of an ecosystem.
(ii) Natural resources are always available in a finite quantity.
(iii) Renewable energy sources should be developed and used.
(iv) Future generations also have as much right to enjoy the environmental resources as the present population.
(v) Environment has a limited self purification capacity.
(vi) Pollutants are likely to travel long distances and have more residence time through land, water or air.
(vii) Quality of life is improved if air, water and land environment is not contaminated with waste products.
(viii) Pollution is a crime punishable under the Indian Penal Code.
(ix) Environmental protection should be an integral part of all developmental activities.

For a safer environment, pollution must be controlled by the following measures:

- Monitoring the environment for pollution levels at regular intervals
- Minimizing the generation of wastes and purifying them before disposal
- Recovering and recycling the useful materials to the extent possible
- Prohibiting the use of highly toxic chemicals
- Strictly adhering to the safety practices
- Implementing an environmental pollution policy for sustainable development

8.5 POLICIES FOR QUALITY IMPROVEMENT

Problems

The present environmental problems may be mainly due to the following:

(i) Rapid increase in population growth

(ii) Indiscriminate disposal of wastes without purifying to the extent required
(iii) Deforestation, desertification, lowering of ground water table
(iv) Release of toxic chemicals into the atmosphere without any control at source
(v) Absence of implementation of pollution control measures or any other regulatory mechanism
(vi) Unavailability of the technology in some cases to treat wastes economically
(vii) Ignorance, negligence, lack of ethics and accountability

Wealth and resources are not uniformly distributed and the developing countries also have to face the consequences from developed countries, in addition to their own deficiencies. It is difficult to understand why people are facing the problems when the environmental quality can be monitored accurately, effluent quality standards are prescribed, treatment process technology is available and when regulatory clauses are incorporated in legal provisions? In some cases, when the consumer is aware of dangers due to pollution, public agitations are stopping the development process in all respects. There must be a policy for the government (Ministry of Environment and Forests) and the pollution control boards should implement them at all costs.

Policy

The proposals from various departments may be listed as follows:

(a) Department of Environment and Forests
   (i) Conservation of resources, flora, fauna, forests and wildlife
   (ii) Pollution control and protection of the environment
   (iii) Sustainable development

(b) Department of Agriculture
   (i) Promotion of organic farming, use of bio fertilizers and pesticides
   (ii) No diversion of agricultural land for other purposes
   (iii) To encourage proper crop patterns compatible

(c) Irrigation Department
   (i) Small projects and optimal utilization of water resources
   (ii) Watershed management
   (iii) To promote adequate drainage facilities

(d) Department of Forests
   (i) To preserve natural heritage
   (ii) To increase forest productivity
   (iii) To prevent illegal use of forest products

(e) Pollution Control Board
   (i) To provide sufficient knowledge and training to different people in all fields
   (ii) To permit the plans for industrial growth
   (iii) To implement the legal provisions on environmental protection.

On all these issues, there must be an environmental management plan for each industry or organization. Periodical energy audit, safety audit and environmental audit helps an industry to maintain standards in accordance with international organizations. Conservation programmes are benefited by recovery, reuse and recycling resources to the extent possible. Public participation, media cooperation and non-governmental organizations act as task forces to safeguard the interests at local levels. A number of cleaner technologies have been developed to improve various industrial processes for minimizing the emission of pollutants.

Environmental education from KG to PG studies since 20 years has provided adequate awareness so that environment can be protected by all the concerned.

International level meetings

United Nations environmental programmes have started with various conferences on different aspects of pollution, depletion of ozone layer, global warming etc. for more than 35 years. A brief list is presented here.

1972–Stockholm conference

Industrialization, causes for pollution and threat to environment.

1982, 1987 Kenya (Nairobi)

World commission on environment and development taking into account socio-economic factors and sustainable environment.

Montreal protocol 1987

Protection of ozone layer in stratosphere and phasing out the production of green house gases.

Rio (Brazil) 1992, 1997

Poverty eradication, water supply and sanitation, health, protection of natural resources and biodiversity, alternate energy systems, sustainable development, climate changes, polluter pays for prevention and protection of the environment.

Cairo (Egypt) 1994

UN conference on demography and development, which includes family and child welfare, female literacy and population control policies.

Kyoto protocol (Japan) 1997

Fight against desertification and drought conditions, reduction of greenhouse gases.

Johannesburg 2002

Poverty and population explosion are the main causes of pollution.

Most of the developed and developing countries have participated in these meets and have resolved to reduce pollutant levels in the human environment with the main objective to provide quality life.

It must be the policy of every organization or management to use environment friendly technology, to discharge the effluents depending upon the prescribed limits only and monitor the quality of environment at regular intervals.

Human rights include right to life, liberty, property and security of an individual. Freedom, justice and peace should prevail.

Intellectual property rights outlines that the rights of material developed by an intellectual should rest with that particular individual and should be protected by the state. The purpose is to protect from exploitation.

All policies must be towards development of cleaner technologies and environmental strategies to reduce the risk to human life.

SUMMARY

EIA is a study to understand the adverse effects due to implementing a project like construction of a dam or starting a thermal plant on the surrounding environment, so that they may be suitably rectified. Environmental audit is a procedure to internally evaluate the various stages to minimize waste production, recovery and recycling materials, and purification systems.

Environmental legislations help to improve the natural environment including forests, lakes, wildlife, etc. It is the duty of the pollution control boards under the department of ministry of forests and environment to save the environment from the adverse effects of pollution.

KEYWORDS

Audit, environmental impact assessment, environmental laws, management plan, pollution control board, recycling of wastes.

SOLVED QUESTIONS

1. **Explain the methodology to arrive water quality index (WQI).**

   To find water quality index, given the concentrations of fluorides as 6.5 mg/L (permissible level 1.5 mg/L only) and chlorides as 1200 mg/L (permissible level 250 mg/L):

$$\text{WQI} = \frac{\Sigma\,(\text{Ratings} \times \text{Weightage factors})}{\Sigma\ \text{weightages}}$$

Ratings	*for fluorides*	*for chlorides*
	1.5 to 3 mg/L – 1	250 to 500 mg/L – 1
	3 to 6 mg/ L – 2	500 to 1000 mg/L – 2
	6 to 9 mg/ L – 4	1000 to 1500 mg/L – 4
Weightages	4	1

$$\text{WQI} = \frac{(4 \times 4) + (1 \times 4)}{4 + 1} = 4$$

(*Note:* The quality is good if it is nearer to 1)

2. What are the main objectives of Environmental Acts?

- To preserve and protect all the environmental assets and their quality
- To prevent the disposal of wastes into the atmosphere without adequate treatment.
- To punish the guilty persons or industry for the failure to comply with the directions of Pollution Control Board

3. State some difficulties encountered in the application of environmental laws.

- Poverty, illiteracy and lack of social discipline
- Difficult to establish cause and effect relationship
- Inadequate technologies either to quantify pollution level or to completely remove the pollutants to the required level
- No separate environmental courts for effective disposal of cases and grant remedies

4. Explain the concept of a common effluent treatment plant for a group of industries.

Liquid effluents from a group of pharmaceutical or chemical industries (located at one place in an industrial estate or nearby) can be collected and treated in one treatment plant effectively. After treatment, wastes can be disposed into a municipal sewer. Industries may have primary treatment plants for treating effluents within their premises to some extent advantageously.

(i) Acid and alkaline wastes may be mixed to obtain the advantage of neutralization.

(ii) Strong and dilute wastes may be mixed to reduce the strength of effluents.

(iii) The entire treatment can be done under the supervision of one expert team and the finances can be shared by different industries.

(iv) For wastes having similar characteristics (homogenous in nature), treatment efficiency is very good.

5. Cite the instances where public participation helped in the environmental management.

(i) Water harvesting in Laporia (Rajasthan) and watershed management in Rallegalsiddi (Maharastra) improved water resources significantly in the village.
(ii) Revegetation of hills in Sukhomagic village in Shivalik hill range protected mountain slopes and landslides.
(iii) Extraction of radioactive minerals near Miriyalaguda (A.P.) has to be stopped temporarily due to public participation.
(iv) Villages near Patancheru (A.P.) got water supply through tankers as the land and water resources are polluted.

6. Give some useful hints to protect environmental quality.

(i) Adopt cleaner technologies.
(ii) Reduce or substitute chemicals which are safer.
(iii) Monitor environment quality regularly.
(iv) Reuse the material to the extent possible.
(v) Provide treatments in all industrial units.
(vi) Dispose effluents only when the norms are satisfied.

7. Try to network the impact of tourism on a lake.

This type of networking [Figure 8.3] helps in knowing the weak spots and manage them while improving tourism.

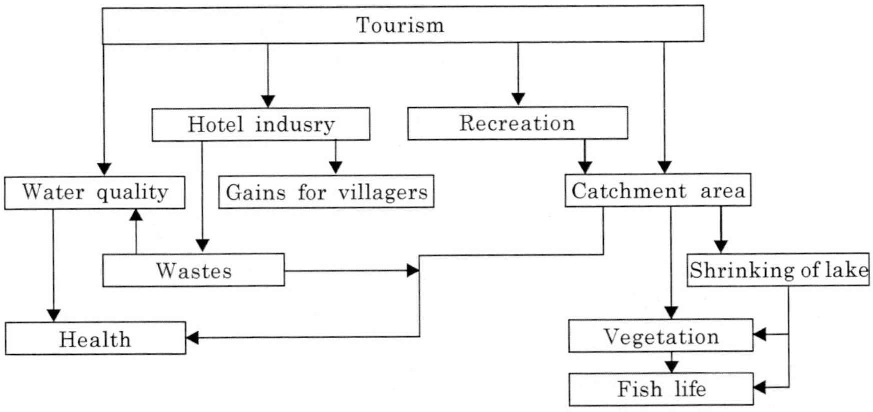

FIGURE 8.3 Network system for tourism project.

8. Waste is a resource. Give two examples in a rural setup.

(i) Composting to obtain good biological fertilizer.
(ii) Fuel pellets from garbage as a substitute for coal.

9. Explain the term 'environment friendly'.

Plastic materials are not bio-degradable. Chemical pesticides reach water and land resources and also humans through agricultural yields. Paper can be recycled and used as a pulp material. That means some products produce negative effects in the environment and some are beneficial to the environment. organic farming is eco-friendly.

10. What is environmental activism?

Sometimes, environment is polluted due to various actions such as (i) disposal of wastes without treatment, (ii) excessive fluorides in drinking waters in some localities, (iii) construction of development projects without caring for the environmental issues.

In such cases, these problems are focussed by the public, known as activism. This causes remedial actions by the government and regulating bodies. Public participation is very much needed in all environmental management programmes.

REVIEW QUESTIONS

1. List two goals of environmental management.
2. Briefly outline activities of irrigation project that disturb the conditions in the sustainable environment.
3. What are the major functions of an environmental audit?
4. Write a note on the constitution of a pollution control board.
5. Explain the principle 'Polluter pays'.
6. Mention the duties of an environmental manager of an industry.

OBJECTIVE TYPE QUESTIONS

Fill in the blanks with suitable words:

1. ___________ day is celebrated to promote environmental awareness.
2. An ozone hole was discovered by satellite ____________.
3. Bhopal gas tragedy occurred in _________ industry.

   (a) fertilizer (b) nuclear power (c) pesticide (d) battery plant
4. Fly ash is a waste from __________ industry.

   (a) power plant (b) cement plant (c) steel plant (d) none of the three
5. Right to life is guaranteed under _________________.
6. ______________ prescribes the standards of water quality from time to time.
7. Public opinion is considered in problems related to environmental quality. ________. (Yes/No)
8. Definition of pollutants is needed for understanding the provisions of ________.
9. Environmental audit is a pre requisite before setting up an industry _____. (Yes/No)

CHAPTER 9

SUSTAINABLE DEVELOPMENT

TOPICS COVERED

- Ethics
- Environmental Stress
- Sustainability
- Self-purification and Regeneration
- Action Plan
- Computerization and Information Technology

9.1 ETHICS

Human nature is bestowed with a number of good qualities like courage, devotion, sincerity, honesty, purity, patience, love, kindness, character and decision-making mind. However, people develop some special characteristics for the sake of material comforts leading to

(a) Corruption and dishonesty
(b) Irresponsibility and careless attitude; use and throw habits
(c) Greed, selfish and commercial interests
(d) No commitment to the society and no respect for values

In general, people think that material resources in the environment are common properties and anybody is free to abuse to any extent, expecting others to purify the resources. All things are viewed in terms of financial returns only. Probably, there is no equitable punishment or justifiable reforms for any crime. May be, rules are for good citizens and polluters escape the clutches of law.

Laws of Nature

The following facts have been generally agreed by wise men.

(a) Matter can neither be created nor destroyed. Environmental resources are limited. All things change in nature with time and may appear in different forms. Does it mean creation comes after destruction or along with destruction? Whether progress (development) is possible without pollution (destruction)? Or changes occur only after reaching chaotic conditions?

(b) Diversity is an asset and beauty of the universe. No two objects or minds are similar. Islands, hill ranges, dense forests, variety of flora and fauna, under water marine life, corals, mangroves, sandy beaches, flowering plants, ferns, mammals, birds, reptiles, amphibians, fishes, coelenterates, crustaceans, moluscs, insects, whales, turtles etc. live together as interconnected and interdependent. Man lives in the biosphere, with other biotic and abiotic systems. Competition, exploitation, parasitism, predation, cannibalism and amensalism are common between species. In the struggle for existence, the fittest survives. Some others adjust to the prevailing conditions and live. A symbiotic relationship is beneficial to all the groups of organisms living together. Resources also get regenerated in course of time. The whole system works in a methodic and principled routine with least effort. This is the principle of coexistence.

(c) Nothing happens without a cause. Sometimes, some actions may not be understood by common people. Each and every action causes multiple reactions and even small things create disasters. Depletion of ground water levels may cause sinking of land, climatic changes and may even lead to desertification.

(d) Law of average is only a method of representation. There is no guarantee for the occurrence of an event at a given place and time. Total amount of rainfall may be same every year in a state. But there may be floods at some places and droughts at other places.

(e) Every problem has a reasonable solution, probably it may not be known with the available expertise. But evolution is a continuous process of advancement. Removal of coloured liquid industrial wastes may be costly and difficult now. Biodegradability of plastic wastes seems to be impossible. Working of a common effluent treatment plant may not be effective today. A positive mind may find answers in future for some difficult environmental issues.

Progress

Man uses the environmental resources for his survival, comfort and commercial activities. Giving and taking are two balancing equations within the different groups.

Green revolution, industrialization, urbanization, globalization and privatization, all these have led to changes in life styles, infrastructure development, material comforts and commercialism. Colleges, cars, cell phones, hotels, hospitals etc. are in one way the indicators of development. Standards of living are improving day by day. Populations, resources and wealth are not evenly distributed. The gap between the rich and the poor is also widening. Population densities in Singapore, South Korea, Netherlands, Belgium and Japan are more but their per capita income is 10 times more than that of India. North Korea, Kenya, Myanmmar, Nicaragua and Sudan have less density of population and their per capita income is also very less. Thus it appears that there is no direct relation between population growth and prosperity of a nation. But living dangerously leads to depletion and degradation of all resources. Ever increasing population growth rate, poverty, illiteracy and imbalances contribute to different types of pollution levels. Some problems are as follows:

(i) Deforestation
(ii) Soil infertility
(iii) Loss of irrigation land
(iv) Exploitation of mineral resources
(v) Domestic and industrial wastes disposal
(vi) Unhygienic and insanitary conditions
(vii) Changes in climate
(viii) Energy crisis
(ix) Inadequate resources

Whether progress and pollution are inseparable? Whether moral values prevent environmental disasters? Ethics are for practice and not for preaching.

9.2 ENVIRONMENTAL STRESS

Carrying capacity of the environment (i.e. to support maximum number of living organisms in a given time and place) is limited. It is said to be stable only if the utilization of resources does not exceed the rate of replenishment or availability. The capacity for regeneration (self purification) is also limited. Pollutants should be discharged after proper treatment into the environment without affecting the recharging or renewable activities of natural ecosystems.

Just like cough, pain, fever are symptoms of ill health, environment also indicates a number of warning signals of stress development. Some are listed below:

- Discolouration and premature dropping of leaves leading to plant decay
- Colour and turbid drinking water sources
- Lowering of ground water levels
- Algal growth in water sources like lakes (due to excess of N and P)
- Poor visibility due to dust and smoke in the atmosphere
- Oil and floating matter in water bodies

- Stagnation of water and open drains
- Presence of communicable diseases (water, air, insect borne)
- Dissolved oxygen depletion in rivers (due to decomposition of organic wastes)
- Softening of bones of human beings and cattle (due to fluorides)
- Discolouration and disintegration of materials (due to acid fumes)
- Extinction of species
- Unauthorized dumping of garbage in residential areas
- Infertile top soil (due to water logging, salinity, erosion etc.)
- Lack of fresh oxygen in the breathing environment (due to vehicular exhaust)
- Loss of land for agricultural production (due to urbanization)
- Indiscriminate disposal of effluents from industrial units
- Reduction of crop yield
- More green house gases like CO_2 in the atmosphere
- Open mining units
- Presence of toxic chemicals like Pb, Hg, As, CN in the human environment

Unequal distribution of wealth and utilization of resources lead to imbalances in the society. These inequalities may ultimately lead to crimes or migration tendencies in population.

It is always wise to notice the warning signals given by the surrounding environment and prevent further deterioration, otherwise once the land or river is polluted further and further, recovery is almost impossible on the face of earth. Once lost is lost for ever.

9.3 SUSTAINABILITY

People worship land, river, animal, fire and nature as God, as living process is linked with the environment. Survival of human beings depend upon the resources in the nature. Man and environment should live in win-win relationship. Nature is an educational institution for man. We need development without destruction, progress without pollution and resources to sustain future needs also. Resources are utilized by populations. More populations mean more consumption of resources and more contribution of wastes. These effluents degrade the quality of resources and affect the quality of life. When resources are depleted and possess impurities, populations starve and decrease in numbers. A number of species of plants and animals disappear and even in global climate changes take place. Through these chaotic circumstances, populations start their life from scratch, once again. This is the way how things happen in the nature, if mankind forgets their responsibility for a sustained livelihood.

Another important observation is that diverse communities are functionally more producing and stable under environmental stress. Take for example, some tigers, goats and grasslands co-exist. If tiger population improves, number of goats will decrease and grasslands increase. In no time, with inadequate food quantity, tiger population decreases and again goat population increases. Overgrazing occurs when the goat population is high. This will be a chain reaction in course of time. If deers, buffaloes, lions, tigers and a number of other species co-exist in a given space and time, there will be no shortage of food at all. Just like five fingers of a hand or different parts of the body doing different functions, diversity is essential for sustenance and stability.

One opinion is that pollution is due to prevailing socio-economic conditions. Consumptions of energy and other resources will be mainly utilized by the twenty per cent rich population on the per capita basis. However the majority of populations living below poverty line also consume a major share of resources on the whole. Removal of poverty, reduction of population growth rate and improving educational values may take time, but creating awareness towards the sustainable development should take its priority.

River valley projects, mining, urbanization, industrialization and commercial interests, all are leading to social development on one side and exploitation of resources on another side. The relationship between demand and supply, material utilization and waste generation and progress and pollution appears to be very fragile in nature. Whether sustainable development and high energy consumption go together? There must be a balance between conflicting events and actions to establish a positive relationship between man and biosphere. There is nothing like waste in the environment. Waste is also a resource, if properly used or technologically disintegrated into useful forms. Wisdom should play a major role for the benefit of mankind and the total living habitat. That is the art of living!

Some activities for the sustainable development are given below:

- Conserve, preserve and protect historic, heritage, cultural monuments and locations.
- Prevent wastage of food products.
- Recover, recycle and reuse materials like paper, metals.
- Avoid and minimize loss, breakage or wastage.
- Save energy.
- Develop alternate sources of energy.
- Store for the future generations and needs.
- Encourage eco-friendly products.
- Allow the resources to replenish and regenerate by natural means.
- Grow trees.
- Ban roads and temples into deep forest locations.
- Prefer mass transportation systems to individual vehicles.
- Supplement, compliment and support social welfare programmes.
- Adopt simple living.
- Efficiently manage the entire system.

Some resources (trees) will have monetary value and others (sea) have existence value. Some are renewable. Some need periodical maintenance. There are special programmes taken up by the Government. Some are given below.

(i) River Ganga purification action plan
(ii) Removal of sediments from lakes
(iii) Supplying surface water to fluoride affected areas
(iv) Interlinking of rivers to prevent floods and droughts
(v) Afforestation, social forestry, farm forestry, use of bio-fertilizers and integrated pest management
(vi) Wildlife conservation projects for tiger, elephant, crocodiles etc.
(vii) Conservation of wet lands at number of places for example, Chilka, Sambhar, Wulbar, Loktak, Harike
(viii) Development of national parks at Kaziranga, Manas, Kaoladeoghana, Nandadevi, Sunderbans
(ix) Watershed management and water harvesting methods

All these programmes are intended for sustainable development and to maintain a symbiotic relationship between man and his surrounding environment.

9.4 SELF-PURIFICATION AND REGENERATION

Environment is bestowed with a divine characteristic of self-purification and natural recycling process. The six essential elements i.e. carbon, nitrogen, sulphur, phosphorous, hydrogen and oxygen are regenerated in a systematic biogeochemical natural cycle. The same is true for water in hydrological cycle. Food is also converted into waste matter and again re-circulated through land, water and plant life. The human environment is possessing thus a carrying capacity. That capacity is in fact the energy in the balancing system. All eco-systems depend upon the supporting energy for regeneration purposes. Each and every system will have a threshold limit after which the system stops functioning. When a system reaches a tolerance level, it means the assimilative capacity is exhausted. Depletion and degradation of the environment is due to misuse and exploitation of resources. That is, the rate of regeneration is far less in comparison with the rate of degeneration. Ultimately the carrying capacity or the sustainability of the environment is affected.

The second aspect is that the nature allows itself to purify and reach the original, normal, and safe position with a set time frame and spatial requirements.

Harmful pollutants are changed into different compounds which are beneficial to the environment. The various self-purification processes can be mentioned as follows:

(i) Pollutants are mixed, diluted and dispersed in water bodies or in the ambient atmosphere. Strength is reduced and pollutants are carried away from the source of generation of waste.
(ii) Organic matter is decomposed or digested into stable compounds by the action of microorganisms.
(iii) Oxidation and reduction change the state of chemical compounds.
(iv) Solids deposit (movement in the downward direction) or float (movement in the upward direction) depending upon the local situation.

(v) Evaporation or drying is due to natural sunshine in the area.
(vi) Land acts as a filter to separate solids from liquids.
(vii) Absorption, adsorption, combustion, and condensation are also the processes that help in the self-purification process in the natural environment.

Sustainability of an environment depends on the capacity to regenerate and recycle in nature. An index can be formulated on the basis of carrying capacity of the environment for sustainable development for different countries or different regions.

9.5 ACTION PLAN

(i) At International Level

United Nations under development programmes on environment related issues, conducted a number of conferences during 1972 (Stockholm), 1975 (Belgrade), 1985 (Vienna), 1987 (Norway), 1989 (Basle), 1992 (Rio). Their discussions and resolutions were on different subjects like biodiversity, desertification, ozone depletion, climate change, disposal of hazardous wastes, health of oceans, cleaner technologies, sustainable development and environmental education. Certain decisions on global issues are as follows:

- To ban the use of CFCs which cause ozone layer thinning in the stratosphere
- To reduce gaseous emissions like SO_2, NO_2 for preventing acid rains
- To emphasize afforestation programmes as forest balances an ecosystem
- To apply all safety measures while disposing hazardous wastes
- To conserve biodiversity and protect the endangered species
- To use organic fertilizers and avoid chemical pesticides
- To safeguard marine ecosystem

(ii) At Government Level

Ministry of Environment and Forests (through Pollution Control Boards) at State and Central Government's level is engaged in various activities to protect the environment such as:

- Conservation of ecosystem
- Water management practices
- Afforestation measures
- Wasteland reclamation
- Monitoring the environment
- Use of non-conventional energy resources
- Eco-friendly technologies
- Pollution control measures
- Environmental policies and laws

- Environmental education
- Public participation and awareness
- Involvement of non-governmental organizations

(iii) Awareness

Environmental protection is a job with multiple functions. In the periods of 1950–1960, civil engineering students used to study water supply and sanitary engineering as a subject. That was the time of early recognition of disease transmission through environmental route. When water supply and drainage projects are implemented in towns, communicable diseases were under control. By 1970, disposal of liquid wastes from domestic and industrial origin created water pollution problems. Chemical processing and microbial decomposition technologies were used to purify waste water. With a lot of industries not observing purification devices, air pollution problems have come up. Urbanization has led to vehicular emissions and solid waste disposal problems. Later, more serious global problems caused concern. Co-ordinated efforts are needed to maintain clean surroundings at all levels. Persons with experience in various fields of agriculture, medicine, chemistry, biology, civil engineering and chemical technologies entered the field to solve different environment problems. Though the subject is of multi-disciplinary nature, this is the age of super specialities. Any individual can take suitable decisions only with proper and adequate environmental awareness. A person interested in foundry or manufacturing technology or thermal power plant should also know the adverse effects on the environment. That is the main reason for the principles of environmental science and engineering to become an integral part of all education schemes.

If cosmetics can cause cancer, mobile phones infertility, cars pollute streets, water carries disease causing germs and plastics have no place for disposal, what is the end product of all the so-called developmental activities? Nobody has a right to create problems to others. Environment does not have boundary limits. It is time to wake up now and save our divine environment on a priority basis from all corners.

9.6 COMPUTERIZATION AND INFORMATION TECHNOLOGY

After agricultural and industrial revolutions in the world, computers and communication networks have created another modern technological advancement. Accessibility to any information from anywhere on the globe has given scope for assimilation of knowledge. It may be on population forecasts, variations in the climate, availability of resources or pollution levels. Further a realistic view of social structure and status of a place can be obtained from the database of population growth rate, age structure, male to female ratio, morbidity and mortality rates, literacy and income levels and the infrastructural facilities available. On the basis of this information, policy decisions can be visualized, planned and implemented to improve the social environment. Similarly, continuous development of database on water quality and quantity, wind quality and movements, land use patterns, extraction of mineral resources, consumption patterns of energy, helps in solving problems with regard to the depletion and degradation of resources. Repeated procedural calculations and complex designs associated with environmental problems like design of water mains and sewer networks, monitoring environmental quality and purification of solid, liquid or gaseous effluents can be solved with ease, accuracy and rapidity. Information technology also helps to predict the future global or local trends, to prevent environmental disasters and to find out suitable need and time-based solutions for the improvement of quality of life on earth.

SUMMARY

Environmental resources and its carrying capacity are limited. Utilization of resources should not exceed the rate of replenishment. Man lives in the biosphere together with other biotic and abiotic systems and contributes to different types and levels of pollution. Survival of mankind depends upon the strength of resources. Care should be taken to have a balance between conflicting events and acts (such as progress versus pollution), to finally establish a positive relationship between man and the biosphere. Sustainability means to conserve, preserve and protect the environmental resources for the needs of future while adequately managing for the present welfare and development.

KEYWORDS

Environmental ethics, environmental stress, laws of nature, nature's self purification property, sustainable development.

SOLVED QUESTIONS

1. **What are the problems associated with rural environment?**

   They do not have treated water supplies, latrines, road connectivity, medical and educational facilities.

2. **Suggest some projects for rural development.**

   (i) Water harvesting and watershed management
   (ii) Vermi composting
   (iii) Biogas plants
   (iv) Solar energy utilization

3. **How is it possible to quantify 'sustainable development'?**

   Environmental sustainability can be estimated through a number of parameters such as

(i)	Resources (forests, energy etc.):	Availability, utilization, regeneration, management, conservation, depletion, degradation etc.
(ii)	Progress indicators:	Per capita income, standards of living, facilities available, industries, agricultural productivity, health care, literacy levels, population welfare measures etc.
(iii)	Pollution levels:	Waste minimization, reuse, waste treatment and disposal, air/water quality, soil fertility etc.
(iv)	Social conditions:	Moral values, unemployment, crime rate, women and child welfare etc.
(v)	Others:	Animal welfare, livestock, aquatic life, wild life sanctuaries, national parks, forest cover etc.

Data can be collected on each parameter to assess the carrying capacity of the environment. An attempt can be made for a logical summation of various items to obtain a reasonable quantification of environmental assets.

4. Mention the steps necessary to protect the environment.

The Reasonable measures are as follows:

- Creating awareness and public participation
- Developing technology, training and research
- Improving the living standards
- Controlling population explosion
- Following legislative measures
- Giving incentives
- Imposing penalties
- Curbing commercial exploitation

REVIEW QUESTIONS

1. What do you understand by the terms progress and development?
2. Explain some features of environmental stress.
3. Write a detailed note on any two programmes of sustainable development.
4. What is the role played by State Government in the environmental welfare programmes?
5. Do you think moral values prevent pollution in the environment? Explain your views.

OBJECTIVE TYPE QUESTIONS

Fill in the blanks with suitable words:

1. __________ relationship is beneficial to all the groups of organisms.
2. Ground water depletion indicates __________ .
3. __________ communities are functionally more stable.
4. Preserving resources for the future generations is known as __________
5. Environment does not have __________ boundary limits.

Part Three

ENVIRONMENTAL SCIENCE

CHAPTER 10

ENVIRONMENTAL CHEMISTRY

TOPICS COVERED

- Introduction
- Chemical Cycles in Nature
- Chemical Reactions
- Toxic Chemicals
- Applications in the Treatment Processes

10.1 INTRODUCTION

A number of organisms live on the earth. They need food and energy for their activity. **Oxygen** is essential for all forms of living systems. Initially, algae and small plants released oxygen into the atmosphere through photosynthesis. Oxygen is available in the air and air is in constant circulation. With more availability of oxygen, more plants and animals have grown up. This oxygen has accumulated in the form of ozone in the stratosphere and protected the living organisms by absorbing UV radiation. Oxygen is involved in the processes of respiration, combustion, oxidation and degradation of organics and photosynthesis. A constant exchange (absorption or release) of oxygen keeps the ecosystem functioning, with more sustainability.

Water is also very essential for all living things on the earth. 75 per cent of surface on the earth is covered with water bodies. But only about 1 per cent fresh water is available for human consumption. However, a major portion of water either in the oceans or in the deeper layers below the ground has a non-performance asset value. Large amounts of water, when removed from deeper layers of ground causes subsidence of land. Temperature on the earth is affected by the water bodies. Oceans act as a sink for carbon and other compounds. Agricultural properties of soil depend upon the soil moisture. That is to say, water quantity not participating in the hydrological cycle has an indirect and very significant value in the human environment. Water cycle consists of evaporation from surfaces of the water bodies, transpiration from plants, formation of clouds, precipitation and rainfall, surface run off reaching rivers, lakes and infiltration into layers of ground below. Finally water reaches the ocean or is stored in deeper layers below groundwater table. Water from the top soil is used by plant life and from subsoil layers is used for human activities. Figures 10.1 and 10.2 relate to the hydrological cycle and zones in water bodies.

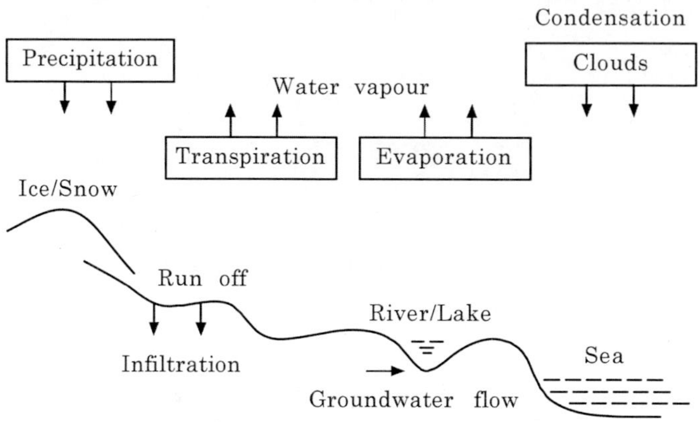

FIGURE 10.1 Water cycle.

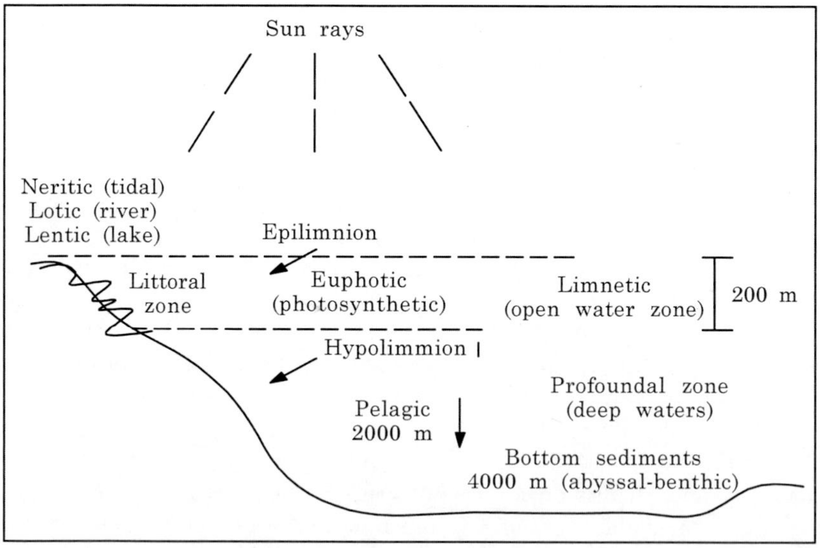

FIGURE 10.2 Zones in water bodies.

Figure 10.3 shows features of the earth from core to crust having geological importance whereas Figure 10.4 presents the soil profile. Soil is formed due to a continuous process known as *weathering*. Wind, water and a

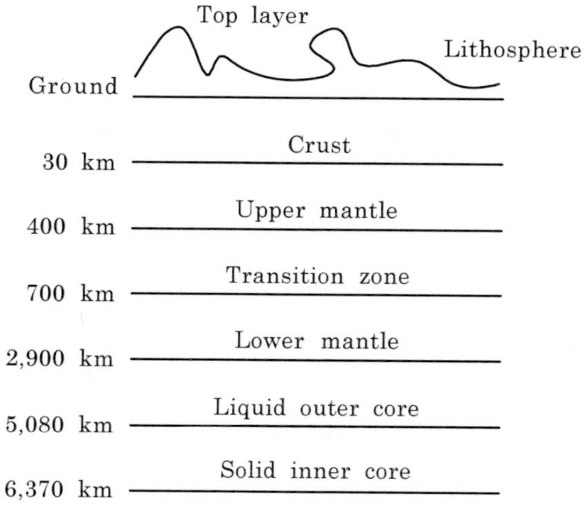

FIGURE 10.3 Features of the earth.

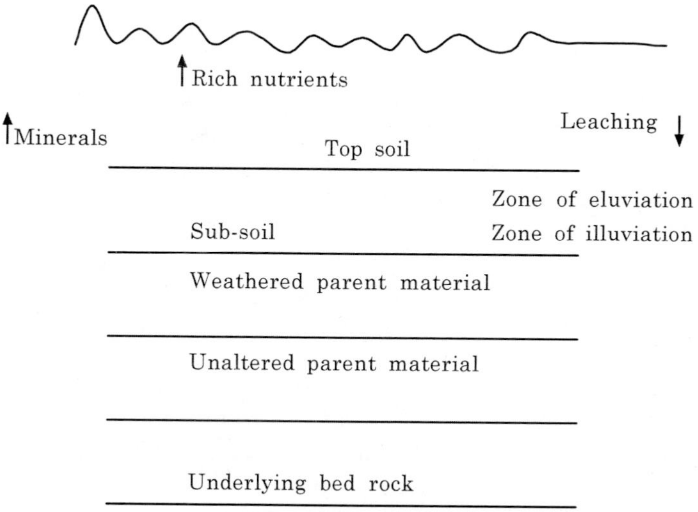

FIGURE 10.4 Soil profile.

number of other agents disintegrate the rock formations and transport the particles for deposition. The nature and properties of top soil is important for agricultural practices and development of vegetation over land surfaces. Fertile soils contain nutrients and minerals and retain the required moisture content. Sub-soil contains water bearing strata. Porosity (the capacity to retain the moisture content) and permeability (the property of allowing water to pass through soil) are the two properties which indicate soil-water-air relationship. Sub-soil also contains coal, petroleum, iron ore, aluminum, phosphates and other resources.

Fossil fuels are the dead remains of plants and animals and the carbon content available over a period of centuries for utilization by the mankind as energy. World can be divided into different kinds of land forms, like mountains, valleys, coastal areas, water falls, lakes, volcanoes, deserts, forest lands which also contain varieties of animals, birds and insects. The types of forests on land vary with altitude and latitude as shown in Figure 10.5. Growth, decay and

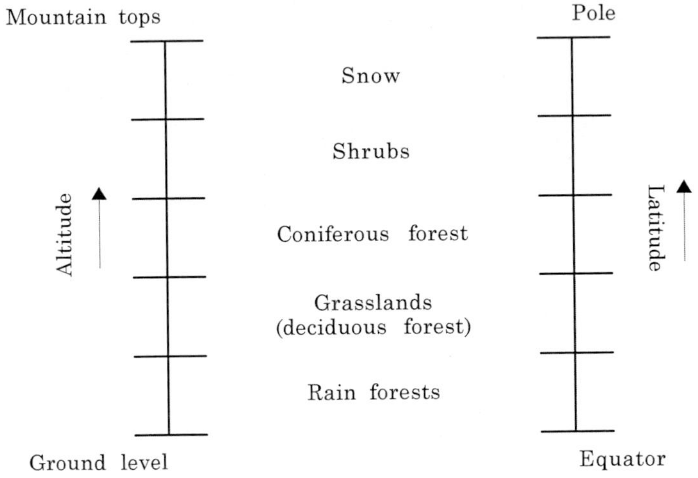

FIGURE 10.5 Forest zones.

death of various plants and animals, is the process of evolution and extinction over a time period depending upon a number of environmental factors. Co-existence of diverging values appears to be the beauty of universe.

Earth absorbs two third of Solar flux, electromagnetic radiation from the Sun and reflects or scatters back one-third of UV (short wave) visible and infrared (long wave) radiation back into the space. 17 per cent of solar radiation is absorbed by the clouds, water vapour, carbon dioxide, directly heating the atmosphere. 5 to 10 square metres of surface area collects 1 kilowatt solar power. The factors which affect the heat balance on the earth are as follows:

(i) Circulation of air mass
(ii) Interaction of molecules and atoms
(iii) Radiation of energy in the infrared (terrestrial) region
(iv) Dust particles in the atmosphere and scattering light
(v) Plants (forests) utilizing carbon dioxide
(vi) Water vapour undergoing evaporation and re-absorption of heat

Deforestation results to less CO_2 in the atmosphere and combustion releases heat and more CO_2. More of evaporation leads to less moisture content in the atmosphere leading to dry conditions. Less evaporation means low rainfall. Forests, oceans and rocks are sink for the regulation of CO_2 in the environment. Soil characteristics, availability of sunlight, intensity and duration of rainfall, forest lands and oceans try to balance the temperature on the earth at a place.

Living environment contains air i.e. 78 per cent Nitrogen, 21 per cent Oxygen, 0.2 per cent water vapour, CO_2 and a number of other gases (by volume). Atmospheric density, pressure and temperature vary with altitude. Figure 10.6 shows the various layers of atmosphere above sea level as datum.

	Thermosphere (hot upper atmosphere)
	Mesopause
80 km	
	Mesosphere
55 km	
	Stratopause
42 km	
	Stratosphere (ozone 5 ppm, temperature increases with height)
17 km	
	Tropopause
10 km	
	Troposphere (temperature decreases with height) 1 km Boundary layer
	Sea level

FIGURE 10.6 Atmospheric layers.

The lower atmosphere is electrically neutral whereas the upper atmosphere is ionized. Pressure drops, temperature falls and density decreases with increasing altitude. Troposphere is the first ten kilometres above the surface of the earth, wherein the temperature decreases with altitude. But in stratosphere temperature increases with altitude and ozone absorbs UV radiation. Unequal heating and cooling on sea and land surfaces, pressure and temperature variations, earth's rotation, frictional and gravity forces, days and nights, seasonal variations, ocean currents, latitude and altitude, rocks and plants, mountains and valleys always keep the air in constant motion in the lower atmospheric level. Stability and turbulance are associated with wind movement. Weather and climate describe the atmospheric conditions (temperature, rainfall, wind movements etc.) at a given place. Thus the environmental study is related with water, oxygen, carbon dioxide, soil, air movement, land features, plants and aquatic life, micro organisms and interconnectivity with human beings. Chemical reactions involved in the environmental set up are discussed in the subsequent sections.

10.2 CHEMICAL CYCLES IN NATURE

Carbon, nitrogen, phosphorous, sulphur, oxygen and water are all necessary for the sustenance of living systems and they are available in the atmosphere. Their occurrence, use and regeneration form a cyclic pattern in nature, the details are already discussed in Chapter 1.

Carbon Cycle

In the presence of sunlight, green plants produce their energy needs (carbohydrates) and liberate molecular oxygen.

$$6CO_2 + 6H_2O \longrightarrow C_6H_{12}O_6 + 6O_2$$

Living organisms return CO_2 into the atmosphere during respiration process. Organic matter is stored as sediments in ocean deposits or as fossil fuels in the deeper layers of earth. Combustion of coal and oil releases carbon dioxide into the atmosphere and forests utilize CO_2. Carbon dioxide forms carbonates and bicarbonates in water. In contact with elements like Ca or Mg, it forms salts as $CaCO_3$ and $MgCO_3$. 100 ppm of CO_2 in the atmosphere is equivalent to 212 giga tons of atmospheric carbon.

Nitrogen Cycle

For protein synthesis, nitrogen is essential. From plants and animal wastes, Nitrogen is converted into ammonia. In the presence of air and nitrite bacteria, ammonia is converted into nitrite. Again nitrites are converted into nitrates. This nitrification process is represented as

$$2NH_3 + 3O_2 \xrightarrow{\text{Nitrosomonas}} 2NO_2 + 2H + 2H_2O$$

$$2NO_2 + O_2 \xrightarrow{\text{Nitrobacter}} 2NO_3$$

$$NH_3 + 2O_2 \longrightarrow NO_3 + H + H_2O$$

Denitrification is a process of reducing nitrates back into nitrites and molecular nitrogen with Pseudomonas bacteria. Plants can utilize nitrogen only in the form of nitrate or ammonia from soil.

Phosphate Cycle

Phosphorous is a scarce but an important constituent of protoplasm. Rocks in soils and sediments in water bodies are sources of phosphates. They are in soluble and insoluble forms. Plants absorb inorganic phosphates from soil and water and convert them into organic phosphates and animals take it from plants. Death, decay and wastes return back phosphates to soil and water run off and sediments finally reach oceans. More phosphates cause more algal blooms in water bodies.

Sulphur Cycle

Sulphur is extracted from iron ore and is oxidized into SO_2 or reduced into H_2S. Combustion of coal with sulphur as impurities also releases SO_2 into the atmosphere. Sulphur dioxide forms sulphuric acid resulting in acid rain or may photochemically react with smog in the atmosphere, both are of environmental concern.

These natural cycles in soil, water and air environment are chemical reactions completed with the help of microbial activities.

10.3 CHEMICAL REACTIONS

Some basic principles involved in different types of chemical reactions are briefly mentioned below.

(i) In chemical reactions, matter can neither be created nor destroyed. A simple example of oxidation of methane shows that 16 g of methane reacts with 64 g of oxygen to result in 44 g of carbon dioxide and 36 g of water.

$$CH_4 + 2O_2 \rightarrow CO_2 + 2H_2O$$

(ii) In a reversible reaction such as

$$A + B \leftrightarrows C + D$$

The reaction proceeds in both directions at the same time. Under equilibrium conditions,

$$\frac{[C]\,[D]}{[A]\,[B]} = K$$

where

K = equilibrium (dissociation or ionization) constant and

[ ] = molar concentration of reactants under equilibrium conditions.

(iii) Where a solute is dissolved in a solvent, the maximum concentration of the substance in solution denotes the solubility in the given conditions. It is a function of pressure, temperature and characteristics of both the chemicals involved. In general, solids are likely to be dissolved more and gases are likely to be expelled with increase in temperature.

(iv) When H_2O dissociates into H^+ and OH^-, the value of K is observed to be 10^{-14} at 25°C. The value of pH of water is taken to be 7 under neutral conditions as $[H^+] = [OH^-]$. If hydrogen ion concentration is greater than 10^{-7}, it is an acidic solution with pH less than 7. If pH is 4, it indicates $[H] = 10^{-4}$ moles/L and $[OH] = 10^{-10}$ moles/L. In the acid base reactions, acid gives a proton, $[H^+]$ and the base gains the same.

Alkalinity is a measure of water's ability to absorb hydrogen ions from an acid solution without a significant change in pH. It is a neutralizing capacity. Carbonates, bicarbonates and hydroxyl group ions, cause alkalinity. Unless pH is extremely high, carbonate concentration is usually negligible in comparison to bicarbonate concentration.

(v) Rate of a chemical reaction depends upon temperature, concentration, composition and characteristics of the reactants. In chemical reactions with a single reaction, the chemical kinetics i.e. the removal rate (dc/dt) depends upon the order of the reaction as follows:

$$\frac{dC}{dt} = -K \quad \text{(zero order reaction)}$$

$$\frac{dC}{dt} = -KC \quad \text{(first order reaction)}$$

$$\frac{dC}{dt} = -KC^2 \quad \text{(second order reaction)}$$

where

K = reaction rate constant and
C = concentration level of the reactant

The negative sign indicates decrease in the concentration

But normally, the chemical reactions may be expected to take place with mixed or multiple substrates.

Temperature affects the rate of reaction given by

$$K_2 = K_1\ \theta^{(T_2 - T_1)}$$

where K_2 and K_1 are reaction rates at temperatures T_2 and T_1 and θ is a temperature coefficient for that particular reaction.

(vi) **Boyle's Law** states that volume of a gas varies inversely with its pressure at constant temperature. **Charle's Law** states that volume of a gas is directly proportional to the absolute temperature of the gas at constant pressure. The combination of these two laws merged to the development of **Universal Gas Constant** for an ideal gas at standard temperature and pressure. Measured gas volumes at any pressure and temperature can be easily converted into any other required conditions.

In a mixture of gases, each gas exerts pressure independently of the other gases and the total pressure is a combination of all the partial pressures of gases in the mixture. If a number of gases are dissolved in a liquid, partial pressures depend upon the composition of liquid. Gases may be separated and collected through distillation and condensation procedures depending upon their vapour pressures. Different gases dissolve in water depending upon the temperature, concentration of other constituents and their saturation values according to **Henry's Law**.

(vii) Hydrocarbons contain carbon and hydrogen with different structures. New organic compounds are formed with replacement of hydrogen atoms. Methane (CH_4), ethane (C_2H_6), propane (C_3H_8), butane (C_4H_{10}), benzene (C_6H_6) are some examples. Organic acids contain COOH group. Hydrocarbons are released into the atmosphere through plants, automobiles, burning of wastes etc. Hydrocarbons gets oxidized and the photochemical reactions release a number of secondary pollutants. Detergents, surfactants and pesticides are very commonly used and found in the atmosphere. Organic matter decomposes into carbon dioxide and water, utilizing oxygen from the atmosphere.

(viii) Atoms contain protons (+ve), electrons (–ve), neutrons, and nucleus. Some atomic nuclei are unstable and during the spontaneous changes that take place within the nucleus, various forms of radioactive emissions are generated.

Cosmic rays from outer space, Plutonium, Uranium, Thorium and Radon from earth's crust, cause radiation hazards to living beings. They induce unstability in chemical bonds and genetic damage with a cancer risk. Fission products from nuclear energy reactors, stay radioactive even for number of decades. Handling these products is very dangerous.

(ix) Nature appears to have a balancing system. Input and output should balance any chemical reaction or any energy transfer system or natural cycles like hydrological cycle. This is subjected to the consumption, accumulation, conversion and losses in the system. Products may change with time or with other conditions. From the material balance equation, process working system can be properly analyzed. It will help in detecting the loopholes in the system. Chemical reactors are designed using mass balance equations and quantity requirements.

10.4 TOXIC CHEMICALS

Environment is a dynamic system, which carries a range of chemical substances released from various material and man-made sources. These chemicals reach human beings and are toxic to their life and surroundings. Pollution is caused due to the presence of one or more contaminants in sufficient concentration, composition and characteristics for some duration. It may be in the form of injury to human, animal and plant life, damage to materials or property or interference with normal activities. The extent of damage depends upon the nature of the pollutant, concentration of pollutant, duration of exposure and individual susceptibility. Other environmental conditions are also important. While exposure to higher concentrations for a short duration causes acute conditions, continuous exposure to lower concentrations also results in chronic diseases. Pollution interferes with food chain, biogeochemical cycles and affects sustainability of life on earth. Some chemicals get accumulated (bio concentration) through fish, plant or food when it reaches the man. They reach the human body through mouth, nose or skin and ultimately lungs, liver, kidneys, brain, blood and the entire system. Some chemicals are carcinogens or alter DNA system in the cell structure.

Some species like methyl mercury, hexavalent chromium, arsenite (inorganic form of arsenic), lead compounds (organic) are more toxic. Some are essential at lower level concentration but are toxic at higher levels, for example zinc, selenium. Molybdenum is essential for plants but toxic to animals. Copper and Maganese are toxic to plant life only. If fluorides are less than 0.8 mg/L in drinking water, dental caries may occur. If fluorides are more than 1.7 mg/L in water, it may react with bones. Aluminium ions are very toxic to fish life in acidic water bodies. Arsenic, lead, cyanide, mercury, cadmium compounds may inhibit the enzyme synthesis and may affect kidneys or brain. DDT attacks the central nervous system. It accumulates and persists on a long-term basis and hence banned from use. Carbon monoxide displaces oxygen from haemoglobin and retards brain activity. Nitrogen oxides act on lung tissues. Sulphur dioxide causes irritation of the respiratory track, bronchitis and pneumonia. Ozone corrodes metals. Sulphur dioxide reacts with marble surfaces. Dust and soot reaches plant leaves and effect transpiration and photosynthesis. Photo chemical smog reduces visibility, cause breathing problems and injurious to plant life. CO_2 causes asphyxiation. Particulate matter accumulates in lungs. CFCs used as coolants in refrigerators and air conditioners, foam products, solvents, aerosol reach and reside for a long time in the atmosphere. Chlorine from $CFCl_3$ (CFC 11), CF_2Cl_2 (CFC 12) reacts with ozone causing ozone depletion in the stratosphere. One molecule of CFC destroys one lakh molecules of ozone. CFCs are more (15000 times) effective than CO_2 in destroying ozone. Global temperature is controlled by carbon dioxide as it absorbs infrared radiation heating the surface of the earth. Acid rains due to SO_2 and NO_2 affect crops and aquatic life. Oil spill reduces photosynthesis, oxygen intake and marine life. Water logging results in less availability of air for roots in soil pores.

All these chemicals are released mainly due to industrialization, urbanization, population increase, agricultural operations, combustion, deforestation, soil erosion, ground water depletion, mining, irrigation projects, disappearance of lakes, conversion of forest lands, anaerobic decomposition of waste matter and more energy consumption. Most of the pollution problems result on the basis of chemical reactions only.

10.5 APPLICATIONS IN THE TREATMENT PROCESSES

A chemical process is a mechanism involving chemical reactions in a reactor. This may result in the release of effluents or removal of impurities. Chemical pollutants in the atmosphere can be quantified from the material balance equations. Some important applications are presented as follows.

(i) pH

- In the process of coagulation (water treatment), alum $\{Al_2(SO_4)_3\}$ works in a slightly alkaline (6.5 to 8 pH) range whereas iron salts

($FeSO_4$, $FeCl_3$) are effective in forming denser flocs in other ranges of pH (5 or 9). That is one reason for the selection of alum for water treatment. Lime or sodium carbonate is added if bicarbonate alkalinity is not sufficient for the pH range.

- In the process of disinfection of water supplies, addition of chlorine results in the process of HOCl and HCl and later HOCl ionizes into H and OCl. As pH of solution increases, more HOCl dissociates into OCl ions, when pH value is less than 5, only elemental chlorine exists. Above pH of 10, mostly OCl remains. Hypochlorite is more powerful in the disinfection process. Formation of HOCl, which is the desired disinfectant depends upon water pH.
- In anaerobic treatment of sewage sludge, formation of organic acids lowers pH value and conversion of these acids into stable end product increases the pH value. Thus the process needs a suitable range of pH for proper digestion of sludge.
- In the ammonia stripping process for the removal of nitrogen, pH is adjusted for the formation of NH_3.
- In waste treatment process, microbial activity is sensitive to pH of the surroundings. The efficiency of chemicals used in the process may also be pH dependent.
- Acidity, alkalinity, electrical conductivity and dissolved salts concentration are different parameters of water quality measurement for study along with pH value.

(ii) Hardness is caused due to Ca^+, Mg^+, HCO_3^-, Cl^- and SO_4^-. Replacement of calcium and magnesium ions with sodium ions (using resins or zeolites) in an ion exchange process, reduces hardness. The resin may be regenerated again by passing through brine solution. The amount of hardness and the chemical requirements can be worked out based on the chemical equations.

(iii) Dissolved salts are removed in water treatment in the reverse osmosis and electrodialysis units, using a membrane of polymeric material and creating suitable conditions for the flow of liquid.

(iv) Corrosion is controlled by (a) removal of CO_2 from water and (b) adjust-ment of pH, if required.

(v) Salinity of water is removed by evaporation and condensation of water vapour or ion exchange process or using membrane techniques.

(vi) Tastes and odours from water are removed by adsorption with activated carbon.

(vii) Iron salts are removed by oxidation and precipitation.

(viii) Fluorides are removed by (a) passing through beds of activated alumina (b) the addition of alum and lime followed by clarification or (c) ion exchange process through special resins.

(ix) Automobile exhaust releases toxic gases like CO, NO_2, HC depending upon the air-fuel ratio, combustion mechanism, ignition temperature and other conditions. Oxygen supply and requirement controls the process of combustion. The quantity of CO_2 liberated from combustion of fossil fuels can be easily evaluated.

(x) Oxygen demand for the degradation of organic matter can be calculated from the COD, BOD values of waste.

(xi) Sulphuric acid smog or photochemical smog and various reactions leading to PAN formulation depends upon the availability of hydrocarbons, oxides of N and S, fog in the atmosphere.

(xii) Particle size, shape, nature and distribution is important in water or waste water treatment and controlling air pollution.

(xiii) Any reactor (for treatment) depends upon material flows, addition of chemicals, fluid detention time, conditions inside the reactor. From the study of mass balance equations, performance of a reactor can be observed.

When river waters are mixed with industrial wastes, and biochemical oxygen demand of waste water takes dissolved oxygen from the river, pollution levels at any distance in a water body can be calculated to obtain the oxygen balance. Similarly, concentration of air pollutants can be known from the dispersion conditions in the air.

Bhopal gas tragedy and Chernobyl disaster are the two examples of leakage of chemical substances affecting large populations and vast areas of land.

CHAPTER 11

ENVIRONMENTAL MICROBIOLOGY

TOPICS COVERED

- Microorganisms
- Environmental Significance
- Biological Decomposition
- Bio-remediation
- Bio-energy from Wastes
- Other Methods

11.1 MICROORGANISMS

The study of microorganisms and their activities are presented in this section which are involved in the:

(i) Source of disease and transmission from person to person
(ii) Nutrient cycles (like carbon, nitrogen) in the ecosystem
(iii) Decomposition of organic matter in the treatment of wastes like waste water, solid wastes
(iv) Purification of water, waste water, solid wastes, and the self purification process in water bodies
(v) Production of useful material in the process of composting, land fill or sludge disposal
(vi) Various aspects of pollution control in the environment

Unicellular microscopic organisms of size less than 1 to 5 microns are known as *bacteria*. They are of different shapes, (spherical, cylindrical or spiral). A typical bacterial cell is shown in Figure 11.1.

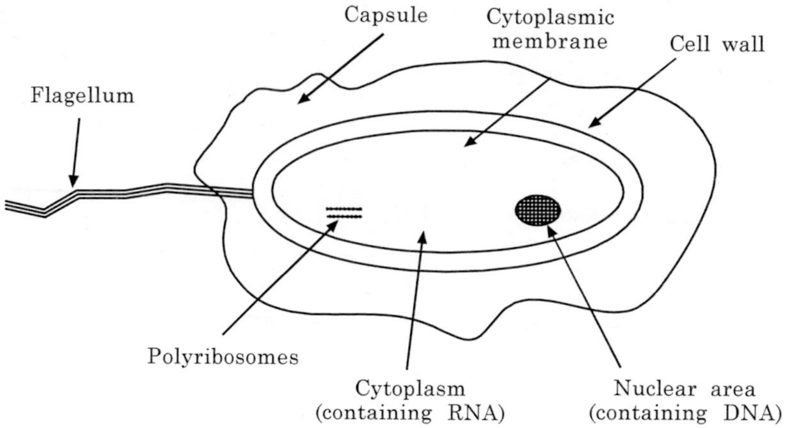

FIGURE 11.1 Bacterial cell.

Microorganisms need the presence of moisture for their growth, a wide range of nutritional needs and favourable environmental conditions. Autotrophic organisms take energy from sunlight or oxidation of inorganic compounds. CO_2 is their carbon source. Heterotrophics take energy and carbon requirements from organic matter. Aerobic bacteria needs oxygen for their growth whereas anaerobic group lives without free oxygen. Facultative bacteria functions with or without oxygen. For most of them, 6.5 to 7.5 is the optimal range of pH value. Bacteria living in the range of temperatures 30 to 40°C are known as *mesophiles* and greater than 40°C are *thermophiles.*

Binary fission (division) is the process of asexual reproduction. They can grow in a wide variety of environmental conditions. Growth rate of bacterial population is directly proportional to the number of bacteria present in the substrate. There are four phases of growth as shown in Figure 11.2. Lag phase is the initial slow period where bacteria needs adjustment with the environment, exponential i.e. very rapid growth rate, takes place under optimal conditions. Due to limitation of nutrients, stationary phase (zero growth) occurs for a

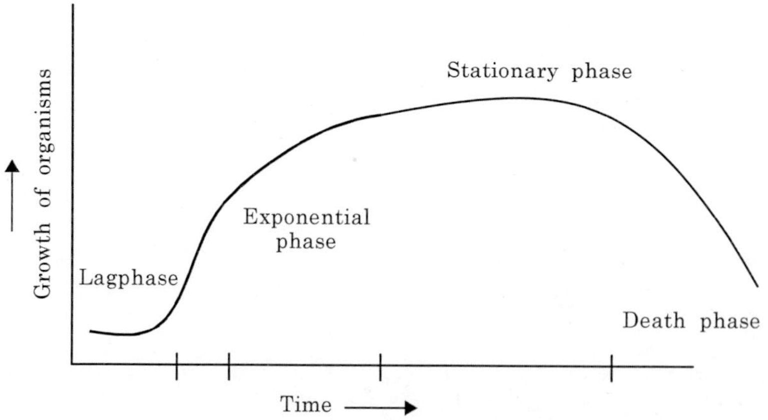

FIGURE 11.2 Typical bacterial growth curve.

sufficiently long period. However later, the death rate exceeds the growth rate due to depletion of nutrients or toxic conditions and declining phase starts.

Viruses which are smaller than bacteria are parasites on the hosts cell and cause a number of diseases like small pox, poliomyelitis, influenza, infective hepatitis.

Algae (green, brown, blue green or other varieties) are small, unicellular or multicellular (of different shapes, with thick cell walls, and with a nucleus), organisms capable of photosynthesis. They impart objectionable tastes and odours to drinking water supplies. Excess algal growth is undesirable as they clog water mains and filters in water treatment units. More chlorine is also needed in disinfection process. However, they supply additional quantity of oxygen in waste treatment plants like oxidation ponds.

Fungi, moulds, yeasts and mushrooms are non-photosynthetic organisms which tolerate varieties of food and other living conditions. Yeasts are useful in the process of fermentation. Fungi are useful in composting of organic wastes. Protozoans (amoeba, sporozoa) are mostly motile unicellular non-photosynthetic organisms present in moist conditions but also survive under adverse situations. *Entamoeba histolytica* causes dysentery in humans. *Plasmodium vivax* causes malaria.

Protozoa and rotifers are present in biological processes in waste water treatment. Rotifers (*Daphnia* and *Cyclops*) are multicellular organisms that feeds on other organisms and organic matter. They have a rigid cell wall and are motile with a tail (flagella). Crustaceans are indicators of unpolluted water bodies. Some of these microorganisms are shown in Figure 11.3.

Microorganisms are denoted by genera and species as *Escherichica coli. Bacillus* and *Clostridium* form spores which are resistant to adverse physical and chemical environment. So they are difficult to be killed by chlorination in the disinfection process.

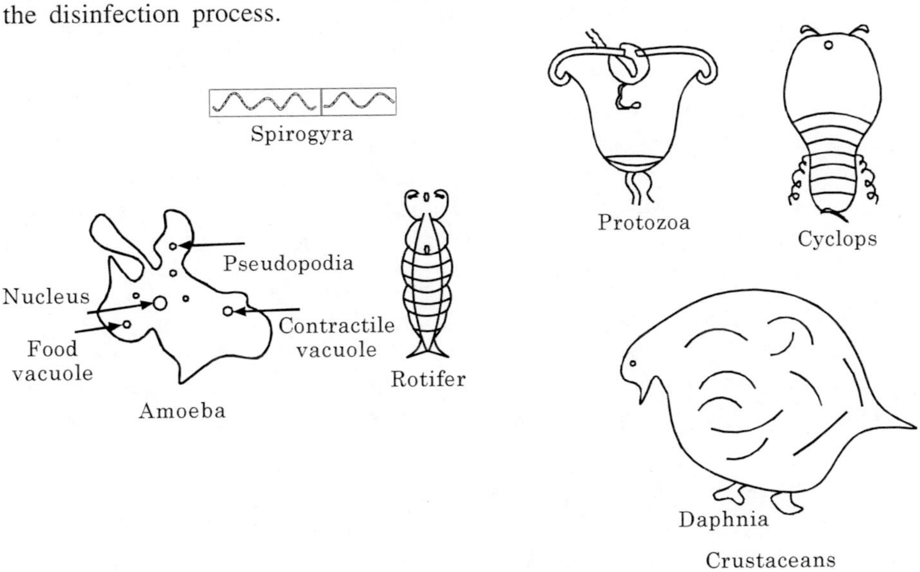

FIGURE 11.3 Microorganisms.

11.2 ENVIRONMENTAL SIGNIFICANCE

Some microorganisms are useful in the natural cycles of nature and decomposition of organic matter but a number of pathogenic bacteria cause diseases in the environment.

(i) In the nitrogen cycle, nitrogen fixation, oxidation and reduction of nitrogen are the result of Azatobacter, Nitrobacter, Nitrosomonas and Pseudomonas microorganisms. Similarly sulphur oxidizing and reducing bacteria, iron oxidizing bacteria are known. Microorganisms are also involved in the process of photosynthesis.

(ii) Algae produce oxygen, taking CO_2 in oxidation ponds or lakes. Bacteria utilize the oxygen for their activity. Algal growth can be observed in elevated water tanks.

(iii) Aerobic bacteria converts organic wastes into stable end products in processes like activated sludge treatment and trickling filters. Protozoa maintain a dynamic balance in the microbial population.

(iv) In the anaerobic process of sludge digestion, organic matter is converted into organic acids by acid producing bacteria, and organic acids are converted into methane by methane producing bacteria.

(v) Microbial activity transforms solid wastes into useful manure in the process of composting. They degrade the complex substances in sanitary land fills also. Decomposing products contain rich nutrient soil with plants or microorganisms to increase the soil fertility.

(vi) Faecal contamination in water may be detected easily by observing coliform group of bacteria. Absence of *E.coli* indicates that water is safe for drinking purposes.

(vii) Microorganisms cause a number of diseases like Anthrax (from animal products), Botulism (through containers), Brucellosis (milk products from infected animals), Cholera (waterborne), Plague and Typhus (through rat fleas), Salmonellosis (food contamination), Shigellosis (waterborne dysentry), Typhoid (waterborne), Hepatitis and Yellow fever (viral infections), Amoeobiasis, Helmenthic diseases, Schistosomiasis, Malaria, Filaria, respiratory diseases due to infected dust, Tuberculosis (*Mycobacterium*). Harmful bacteria enter the human body through mouth, nose, skin or by direct contact.

All these microorganisms can be removed from drinking water by disinfection whereas sterilization ensures 100 per cent kill. Clean environmental surroundings, immunization, and sanitary disposal of wastes are remedial measures for preventing transmission of diseases in the human environment.

11.3 BIOLOGICAL DECOMPOSITION

Pollution brings undesirable change in the environment caused by solid, liquid and gaseous effluents from industries and domestic waste. Carbonaceous and nitrogenous matter exerts oxygen demand depending upon the strength of waste. Complex organic compounds are decomposed and stabilized by aerobic process in activated sludge treatment or by anaerobic process in the sludge digestion units.

$$\text{Organic matter} + O_2 \xrightarrow[\text{Nutrients}]{\text{Cells}} CO_2 + H_2O$$

or

$$\text{Organic matter} \longrightarrow CO_2 + CH_4$$

Any treatment plant gives an opportunity to the microorganisms to convert organic matter and easily separate them from water. For example, an activated sludge treatment plant is a bioreactor. The functioning depends upon the following factors:

(i) Nutrient level in the reactor i.e. food (organic solids) to micro-organisms ratio
(ii) Hydraulic (waste water quantity) loading or BOD loading (input of organic matter)
(iii) Air requirements such as quantity, rate of supply and inlet positions
(iv) Re-circulation of previously oxidized or aerated solids (sludge)
(v) Reaction time in the tank

In a mixing tank, metabolic activities of microorganisms (enzymes, biocatalysts) take place. Protozoans are abundant. It is a homogenous culture under controlled environmental conditions. The efficiency of treatment or BOD removal is very high. Effluent contains stable compounds such as nitrates, CO_2 and water.

Microorganisms can be utilized for the production of biomass and biogas, odour removal, detoxification of spillages, bio-remediation of land sites, biosorption of heavy metals, bio-fertilizers and treatment of toxic wastes.

A reactor unit may be a trickling filter with fixed bed, an activated sludge unit with suspended growth or different unit such as fluidized bed reactor with up flow. Whether a particular waste is amenable for biological treatment or not, has to be judged on the basis of treatability studies.

11.4 BIO-REMEDIATION

Soils are contaminated due to indiscriminate dumping on land sites. These wastelands can be regenerated or recovered by using indigenous microflora or specific microorganisms to cause biodegradation of pollutant. Environmental conditions should be made suitable for the microorganisms to flourish or suitable microorganisms can be selected which can grow under given environmental conditions. Bio-remediation is the process of detoxifying contaminants by the microbial activities. Microorganisms use the contaminants as nutrients or source of energy. If required, additional amounts of nutrients can be added or some of the toxins or predators can be eliminated from the site. Suitable enzymes may be created for degrading the contaminants. This process is very slow with mixed cultures on polluted sites. Aeration of the soil by tilling is helpful. Leachates may be collected so that ground water is not polluted. Sometimes addition of lime and fertilizers help the growth of microorganisms in the soil.

11.5 BIO-ENERGY FROM WASTES

Burning of solid wastes or digestion of organic wastes results in the production of heat energy that can be utilized. Sometimes decomposable organic wastes are also converted for use as manure to the plants.

(a) Combustion of biomass (plants and animal origin excluding fossil fuels) gives energy. Solid wastes, aquatic plants, and agricultural crop residues may be used. The advantages are as follows:
   - (i) Utilization of wastes as a resource
   - (ii) Pollution abatement
   - (iii) Cost effectiveness

(b) Garbage, food products, wastes from dairy or distillery units and other waste material rich in carbohydrates (organic matter) may be digested to produce methane gas. This contains three stages viz., fermentation, acid production and methane production. Different varieties of microorganisms cooperate in the final production of biogas.

(c) Composting is a process of converting organic wastes into rich manure with the help of microorganisms, either aerobic or anaerobic in nature. Earthworms are also used in the process of vermicomposting.

11.6 OTHER METHODS

Biological involvement in a number of environmental engineering processes helps in successfully completing the chemical reactions at a faster rate with stable end products. Some applications of biotechnology are given below.

(a) Biological nitrification is an autotrophic process of oxidation of nitrogen compounds. Algae harvesting removes nitrogen and also heavy metals from water bodies.

(b) Phosphorous compounds are taken up by cell mass in aerobic zone and environmental stress increases the uptake. Phosphates can be removed by a series of anaerobic—aerobic—anaerobic (and sedimentation) processes in waste treatment.

(c) Adsorption of heavy metal by living cells in slimy layers is a biosorption process influenced by a number of environmental factors. Fungi or granular material of biomass are able to grow in the presence of toxic metals and materials. Biosorption is quite useful where physical and chemical processes fail to remove toxic compounds from waste products. It can also be used as a polishing process.

Hazardous wastes need management of environmental conditions such as pH, temperature, nature of microorganisms and wastes. Suitable enzymes can be induced or mixed consortium of species can be used to achieve the results.

(d) Sulphates can be reduced by suitable microorganisms.
(e) Oil film or spills on water bodies may be removed by special types of microorganisms. They help in the breaking of oil films and dispersal or sinking of oil compounds.
(f) Odours can be eliminated by bio-absorption or bio-oxidation (bioscrubber or biofilter) by using Pseudomonas, Mycobacterium or Azatobacter to help chemical oxidation.
(g) Biostimulation or bio-augmentation is the process of adaptation, acclimatization and adjustment for initiating conditions favourable for biological treatment. In future, plastics may be degradable using specific species of microorganisms.

CHAPTER 12

MONITORING AND ANALYSIS OF POLLUTANTS

TOPICS COVERED

- ❑ Sampling
- ❑ General Methods of Analysis
- ❑ Analysis of Water and Waste Water
- ❑ Air Quality Analysis

12.1 SAMPLING

Sampling is a procedure of collecting a representative portion of water, waste water or air from an area to ascertain its quality and characteristics. Samples may be collected from rivers, wells, reservoirs, disposal sites, from ambient air (surrounding human environment), pollution control devices and inside factories.

The main purposes of sampling are the following:

(i) Identify the chemicals and characteristics responsible for pollution
(ii) Know the nature and degree of pollution from various sources
(iii) Design specific plants for treatment
(iv) Monitor the functioning of treatment plants
(v) Ensure the quality of environment as per the standards prescribed by various agencies
(vi) Prevent health hazards

It is essential to know the topography of the area, climatic conditions, land use patterns, location of industries, sources of water and waste water, so that possible types of pollution may be identified.

When the nature and source of pollution is approximately known, one or two litres of random, (instantaneous or 'grab') sample is taken at the time of active release of the pollutants from suitable locations. Composite or integrated samples may also be collected at regular intervals of time or continuously (depending upon the situation) for a specific period of 8 or 24 hours in some cases, to evaluate the average concentrations at the same location over a period of time. Continuous monitoring is very helpful in controlling pollution levels in the surrounding environment.

The following points are to be considered while taking samples:

(i) Sterilized glassware must be used. Care should be taken to avoid contamination from any other sources.
(ii) Samples must be tested immediately, otherwise stored properly to avoid significant changes in the sample.
(iii) The sample should truly represent the characteristics and conditions at site.

12.2 GENERAL METHODS OF ANALYSIS

Volumetric, gravimetric, colourimetric and instrumental procedures are available for the analysis in an environmental science laboratory.

(a) In a volumetric analysis, a fixed quantity of the sample is titrated against known concentration of a chemical solution, under suitable chemical reagents and colour indicators. When the chemical reaction is complete, the total of reactants and products must balance. As the volume and strength of the titrant and the sample volume are known, the unknown parameter concentration in the sample is obtained.
(b) In a gravimetric method, solid matter is separated from water by sedimentation, evaporation, and filtration techniques. The weight of solids per unit volume of fluid is determined.

Colour Comparison Method

Under a set of conditions, the intensity of colour developed is proportional to the concentration of a particular chemical in solution, i.e. the more the concentration, the more the intensity of colour. Relationship between light absorption (a parameter of intensity of colour) through a column length of the sample, at a particular wavelength (depending upon the characteristics of absorbing substances) and concentration of the substance is linear according to **Beer's law**. So, the amount of monochromatic light passing through the sample solution is compared to that of a blank solution. When the transmission through the blank solution is 100 per cent, the percentage transmission of an unknown solution depends upon the concentration of the sample. Hence suitable calibration curves are prepared for different concentrations of chemicals and their light absorbing intensities. An unknown sample is prepared under identical conditions and the concentration is obtained from the absorbance value as per the calibrated curve.

In a visual comparison, colours developed with unknown samples are matched with standard colours developed with known concentrations. Instruments like **spectrophotometers** give the results more accurately. The advantages of colour comparison method, is that a number of samples can be analyzed in a short interval of time.

In a spectrophotometer a lamp emits a broad radiation spectrum in the visible and ultra-violet range and selected wavelengths are isolated and focussed. The intensity of light transmitted by the sample is measured by a photoelectric absorption having a detector and recording system.

Atomic absorption spectrophotometer, is highly useful in measuring the concentration of trace metals in water.

Gas chromatography is used more for analyzing organic pollutants in water or air.

Various parameters like pH, dissolved oxygen, conductivity, turbidity may be analyzed directly using the instruments.

12.3 ANALYSIS OF WATER AND WASTE WATER

Colour

Colour is due to the presence of salts in the solution or in colloidal suspension.

1 mg of platinum salt dissolved in 1 litre of distilled water produces 1 unit of colour on platinum cobalt scale. On that basis, standard solutions of different units of colour are prepared. The colour of a given sample is compared with the standard colours.

Turbidity

The amount of light scattered or absorbed is proportional to the turbidity of the sample.

Standard calibration curve is prepared using hydrazine and hexamine solution, using suitable dilutions. Nephelometer reading with unknown water sample gives turbidity from the calibration graph, in Nephelometer Turbidity Units (NTU).

Total Solids

By filtering known volume of the sample and evaporating the filtrate, the weights of filterable and non-filterable solids are obtained.

The water sample is evaporated in a crucible to dryness in an oven at 103°C till the remaining weight of solid becomes constant. This gives total dissolved solids (mg/L) in the sample.

The above residue is ignited at 600°C in a muffle furnace for about 30 minutes. Solids are weighed after cooling. Loss of weight on ignition gives a measure of volatile solids.

Acidity

2 or 3 drops of methyl orange indicator is added to 50 ml sample and is titrated with 0.02 N NaOH solution till the colour changes to faint orange. Mineral acidity (mg/L as $CaCO_3$) is obtained from the quantity of titrant used.

The titration is continued after addition of 2 or 3 drops of phenolphthalein to a pink end point to obtain CO_2 acidity.

Alkalinity

2 or 3 drops of phenolphthalein indicator is added to a 50 ml sample and is titrated against 0.02N NH_2SO_4 till the pink colour disappears. Phenolphthalein alkalinity (P as mg/L of $CaCO_3$) is obtained from the quantity of titrant used.

The titration is continued after adding 2 or 3 drops of methyl orange till the orange colour turns to pink. Methyl orange alkalinity is known from the quantity of acid consumed in this titration.

Total alkalinity (T) is the sum of the above two.

pH

pH meter is based on the principle that potential developed across a glass membrane due to the concentration gradient between a standard reference solution of Hydrogen ion and sample solution will be proportional to the pH value. Initially pH meter is adjusted with a known buffer solution (4 or 9). Later the electrode is immersed in the unknown sample and the reading directly gives the pH value in the proper range of 0–7 or 7–14.

Colour comparators are used (with a set of universal indicators and standard discs) for different pH ranges to match the developed colour of the unknown sample.

Chlorides

The sample is titrated against 0.0156 N silver nitrate solution till the yellow colour turns red, using potassium chromate as an indicator. Volume of the titrant used is a measure of chloride concentration in the sample taken.

Total Hardness

1 ml of ammonia buffer solution (to maintain pH 9.0) and few drops of Erichrome black T indicator are added to the water sample and is titrated

against standard 0.01M EDTA solution till the colour changes from wine red to blue. Total hardness (mg/L of $CaCO_3$) is obtained from the quantity of titrant used.

Chlorine

A pinch of KI and 10 ml of acetic acid are added to the water sample and is titrated against 0.025 N sodium thiosulphate using starch as an indicator until the blue colour disappears. Available chlorine (mg/L) is known from the quantity of thiosulphate used.

Orthotoludine reagent is added to the water sample and the colour intensity is compared with the standards given, using a chloroscope (colour comparator), to obtain the residual chlorine in water sample.

Dissolved Oxygen (DO)

1–2 ml of $MnSO_4$ and 1–2 ml of alkaline potassium iodide- Azide solution are added to 300 ml sample taken in a BOD bottle. It is thoroughly mixed to dissolve the precipitate by adding 2 ml of conc. H_2SO_4.

200 ml of the above solution is titrated with 0.025 N $Na_2S_2O_3$ using starch indicator until the blue colour disappears.

Dissolved oxygen in the sample (mg/L) is obtained from the quantity of $Na_2S_2O_3$ used in titration.

Dissolved oxygen analyzers may be used directly.

Bio-chemical Oxygen Demand (BOD)

Waste water samples are diluted with (1 litre) distilled water to which 1 ml each of phosphate buffer, $MgSO_4$, calcium and ferric chloride solutions are added.

Dissolved oxygen of the sample is obtained first. These samples are incubated at 20°C for 24 hours (or 5 days).

DO is again determined after the incubation of samples for 24 hours (or 5 days). The difference in the two tests gives the amount of oxygen utilized by organic matter for the given time and temperature.

Strength of the waste is indicated by the BOD.

Chemical Oxygen Demand (COD)

0.4 ml mercuric sulphate, 10 ml $K_2Cr_2O_7$ solution and 30 ml H_2SO_4 are added to 20 ml of waste water sample in a flask.

The sample is refluxed in a condenser for 2 hours. Condensate is cooled and washed with distilled water.

The mixture is diluted to about 140 ml and titrated with standard ferrous ammonium sulphate solution using ferroin indicator, till the colour changes to reddish brown end point.

$$\text{COD (mg/L)} = \frac{\text{Normality of ferrous ammonium sulphate solution} \times 8 \times 1000 \times \text{ml of titrant}}{\text{ml of sample}}$$

Chemical oxygen demand (COD) indicates the chemically oxidizable matter in waste waters.

Iron

A calibration curve for the optical density or (per cent transmission) versus concentration of iron using known samples (adding 2 ml of conc HCl and 5 ml of KCNS to develop colour) is prepared using a spectrophotometer.

For the unknown sample, the optical density is measured and the corresponding concentration obtained from the calibration curve.

Sulphates

For known concentration of potassium sulphate solution, 10 ml of NaCl or HCl, 20 ml of glycerol (alcohol) are added to make up to 100 ml using distilled water and 0.3 gm of barium chloride powder, to obtain turbidity of sample. For different turbidities, a calibration chart is prepared using a Nephelometer.

The absorption of the unknown suspension is measured and the sulphate content is known from the standard chart.

Bacteriological Examination

Water samples in five test tubes (10 ml each) are taken containing nutrient lactose broth. Durham tube is kept inverted in the test tube. All the test tubes are incubated at 37°C for 24 hours. Gas formation in the fermentation tubes indicates the presence of coliform bacteria (positive test) in the given water sample. The test is confirmed after 24 hours of incubation.

From the number of positive findings of coliform group organisms most probable number (using a statistical chart) as a bacterial density is computed in 100 ml of water.

12.4 AIR QUALITY ANALYSIS

Air samples may be taken either from a stack (chimney) or from the ambient atmosphere or from the pollution control devices or from automobiles.

High volume sampler consists of a glass fibre or synthetic filter of (about) 100 sq. cm. exposure area, and pore-opening size of 3 microns. Suction pump draws air sample for a required period of time. Volume of sample of air passing through the filter is measured by rotameter. This sampler is used to collect particulate matter from the ambient air. (Refer Figure 6.3.)

Air is allowed to continuously pass through a sampling train consisting of various bottles (containers) having different absorbents or adsorbents for the study of various gaseous pollutants at a time. (Refer Figure 6.4.)

Suspended Particulate Matter

Particles of less than 10 μ size are collected through glass fibre filters, using a high volume air sampler. Material collected on the filters may be weighed and expressed as mass per unit volume of air drawn.

SO_2

Reaction of SO_2 and lead oxide produces lead sulphate. It is converted to barium sulphate and is estimated gravimetrically or turbidimetrically.

NO_2

NO_2 forms sodium nitrite when absorbed through sodium hydroxide-sodium arsenite solution. Addition of sulfanalic acid and naphthyl ethylene diamine dihydro chloride results in red violet colour whose absorption at 550 nm is measured spectrophotometrically.

CO

Absorbant silica gel impregnated with acid solutions of ammonium molybdate and palladous sulphate is allowed to be in contact with carbon monoxide. The colour intensity varies from yellow to blue green.

Gas chromotography is adopted to analyze CO after conversion to hydrocarbons.

Ammonia

Ammonia develops colour with Nessler's reagent (K_2HgI_4 + KOH), and the colour intensity is matched with standard solutions to obtain the required concentration.

APPENDIX

Some projects of environmental concern are:

1. Interlinking of Indian rivers
2. EIA for a project like Dam construction
3. Selection of site for the disposal of solid wastes using GIS methodology
4. Vermi composting
5. Bio-remediation
6. Energy efficient buildings
7. Development of Rural Environment
8. Draft National Environment Policy (August 2004)

They are briefly explained here.

A.1 INTERLINKING OF INDIAN RIVERS

India has substantial water resources potential. But the availability of water in various parts of the country is highly uneven. Large areas in western, central and southern parts of India receive very low rainfall while the northern and eastern regions receive very high monsoon rains causing heavy floods.

Principal use of water has been for irrigation to achieve self sufficiency in food generation crops. Ever increasing population in the urban areas require water for domestic needs. The concept of transferring water from one river basin for use in another basin is practised in a small scale measure for short distances at various parts of India. Dr. K.L. Rao's **Ganga Cauvery link** proposal during 1970 caught the imagination of people. But the total lift involved was 565 m with a power requirement of 6000 MW for about 150 days. The national perspective plan of 1980 comprised of two main components depending on Himalayan rivers and peninsular rivers, proposing gravity links as far as possible.

Himalayan Component

It envisages construction of storage reservoirs on the main Ganga and Brahmaputra rivers and their principal tributaries in India and Nepal so as to conserve monsoon flows for irrigation and hydro power generation. Inter linking canal systems will be provided to transfer surplus flows of Kosi, Gandak and Ghagra to the west. Surplus flows available on account of interlinking of Ganga and Yamuna are proposed to be transferred to the drought prone areas of Rajasthan and Gujarat. With this proposal about 14 million hectare metres of additional water would be available from these river systems for irrigating an estimated 22 million hectares in the Ganga-Brahmaputra basin, and also provide 1120 cumecs of water to Calcutta. 30,000 kW of hydropower could be generated.

Peninsular River Development

This component consists of four parts, the major point being the diversion of surplus flows of Mahanadi and Godavari to water short river basins of Krishna, Cauvery and Vaigai down south. The other three parts are the diversion of west flowing rivers of Kerala and Karnataka to the east; interlinking of small rivers flowing down along the west coast, north of Mumbai and south of Tapi; and interlinking the southern tributaries of Yamuna. The scheme was envisaged to provide additional irrigation benefits of over 13 million hectares.

Interlinking proposals involve construction of a number of large dams and interlinking canals running to thousands of kilometres, with an estimated cost of about six hundred thousand crores.

Feasibility

The programme needs huge investment and acceptance from different state units (or even groups of people within the state) for sharing surplus water. Construction of large dams submerges vast fertile land areas and habitats. Resettlement and rehabilitation of affected people will be another major problem. Water pollution may be transferred from one river basin to another. Storage reservoirs may cause water logging and changes in climatic conditions. Finally, ecological disturbances are likely to be a major concern.

The solution lies in implementing small schemes first with a proper environmental management plans and adequate compensation packages (financial or otherwise) to the affected people.

A.2 ENVIRONMENTAL IMPACT ASSESSMENT (EIA) OF A DAM PROJECT

EIA is a method of analyzing the possible adverse effects of any proposed project on the human environment. Dam modifies the upstream and downstream river flows. Villages, agriculture lands, buildings, properties, forests and heritage structures may be submerged. Affected people need a resettlement in a new system. Reservoir formation alters the status of the ecosystem. River flow causes erosion at some places and siltation at other places. Deposition of silt reduces the storage capacity of reservoir. Impounding water in large tanks may induce earthquakes in the region. Inundation of land results in the extinction of variety of biological species.

Though dams are beneficial for irrigations of the lands, one should identify, quantify and minimize the evil effects. The purpose of assessment is to find out the ways and means of protecting the human environment from the adverse effects. Unless all the necessary remedial steps are completely assured, the proposed project may not be cleared by the environmental authorities.

EIA requires the following steps:

(i) List out completely the details of the project.
(ii) List out the harmful effects on the environment due to each project activity.
(iii) Suggest suitable remedial or alternate measures to reduce the adverse effects. This can be done by changing the project activities or by implementing environmental protection systems.
(iv) List out the resettlement plans for the affected people.
(v) Finalise the plans that are beneficial in all respects.

A.3 SELECTION OF SITE FOR THE DISPOSAL OF SOLID WASTES USING GIS METHODOLOGY

Generally, low-lying areas and outskirts of towns are used for the purpose of open dumping of solid wastes. They are a source of nuisance value with objectional smoke and odours and serve as a breeding place for flies and mosquitoes. Thus site selection is a management issue based on a number of considerations such as:

- Availability
- Accessibility
- Away but not very far
- Wind direction
- Water/land pollution
- Aesthetics

GIS Application

A number of application software modules are available to solve the problems through integrated multilayered analysis approach. Spatial data in thematic maps and other details are entered into the system, digitized, scanned, and converted into master maps as per the user's requirement. Based upon the significance and relative importance of parameters, different sites under consideration are compared before making a decision. The primary factors which contribute significantly to the site selection criteria include the terrain information, information about the hydro-geological parameters, land use/land cover information, the economic viability of the proposed site, physical feasibility (presence of water bodies, proximity to roads etc.) and factors based on legal, social or political restrictions.

If the site is near water bodies or on sloping grounds, it has to be excluded for obvious reasons. Soil characteristics, ground water table, nearness to the town and all other parameters may be given parameter importance weightage factors. The suitability index value for each site can be calculated from the known values of weightages and relative importance of each attribute. The higher the suitability number for a given area, the more suited is the selection of the prospective site for dumping solid wastes.

GIS is advantageous for handling huge volumes of spatial data. The model can integrate the haul distance into the database to find an optimal route to the disposal site from the various collection points in a locality.

A.4 VERMI COMPOSTING

Vermi composting is a simple and natural approach to use earthworms for converting organic (solid) wastes into a high quality manure. Food processing wastes, dairy and agricultural wastes, animal wastes, plant residues and paper products can be safely and effectively disposed in this method. Organic matter is the food for earthworms and microorganisms. They help in decomposing (composting) waste matter. The left over (vermi castings) contains rich minerals, enzymes and fertilizing elements (N, P, K). Vermi castings are useful as soil conditioners and are excellent nutrients to increase crop productivity. Bio-conversion is a rapid and efficient process. Based on the availability, growth rate, adaptability to environmental conditions, efficiency of conversion and other factors, a number of earthworm species are developed and used in vermi composting units.

The process consists of the following:

(i) Collection of waste biomass, shredding, sorting or separating organic waste from other material (like iron, plastics, glass) and keeping in heaps of 1 m height.
(ii) Earthworms are inoculated in sufficient quantity to decompose the material. This process can also be carried underneath the growing plant to promote synergetic action of earthworms, plant and soil bacteria.

The compost plant works as a bioreactor. Moisture content, availability of oxygen, temperature, pH and activation should be optimally maintained to obtain quality manure in a month or two. These castings improve retention of water by soil, nitrogen fixation, mineralization of organic matter and plant growth hormones.

Advantages

(i) Reduction in volume of solid wastes
(ii) Recycling of wastes

(iii) Use of resource for the sustainability
(iv) No foul odours
(v) No skilled labour or operational problems
(vi) Low costs and highly efficient waste treatment

A.5 BIO-REMEDIATION

Solid and liquid wastes with or without adequate treatment are disposed on land. Decomposed waste matter is retained in soil pores and liquid effluents percolate into lower layers and ground water sources. Toxic chemicals contaminate the top soil and render unfit for growing crops. In due course, such type of lands become useless for irrigation or any other purpose. Bio-remediation means the use of biological (microorganisms, microflora) methods for cleaning soil contamination and also ground water pollution. Land farming methods such as ploughing of soil, mixing of soil, rotation of crops are also methods to rejuvenate soil-air-water relationship suitable for crops. The primary pre-requisite is to treat wastes effectively prior to disposal on land to reduce toxic levels. Some methods available for bio-remediation of soils are as follows:

(i) Water sources may be diverted away from contaminant sites by constructing impermeable soil walls.
(ii) Contaminated groundwater may be pumped out for treatment.
(iii) Contaminated solid waste material may be incinerated at high temperatures.
(iv) Chemicals may be injected into the contaminated zone to detoxify or immobilize the waste matter.
(v) Bio-remediation is the process of modifying site conditions by using micro organisms to promote bacterial degradation of organic wastes.
(vi) Silt material from lakes may be dredged out for the treatment on the shore.

Lining with impervious layers, prevents the contaminants to infiltrate into the ground layers. Otherwise leachates should be collected separately. Some times ventilation helps in the escape of trapped gaseous pollutants from the soil pores.

Bio-remediation helps to stimulate rapid degradation of wastes. Micro-organisms and microflora utilize contaminants as a source of food and energy requirements and detoxify its nature. So, favourable conditions must be created for the enhancement of microbial activity, by the

(i) moisture content adjustment
(ii) pH (addition of lime if required) adjustment
(iii) addition of inorganic nutrients
(iv) availability of oxygen
(v) presence of electron acceptors
(vi) mixed cultures for tackling different pollutants
(vii) removal of volatile organics

Bio-remediation is a slow and economical process for removing toxic chemicals. Suitable micro organisms may be used for breaking oil spills in sea water.

A.6 ENERGY EFFICIENT BUILDINGS

Traditional residential buildings incorporated natural means of lighting and ventilation depending upon the climatic conditions and living standards. The salient features are tiled sloping roof, open verandah, 4 m roof height, more than 30 cm wall thickness, larger openings in north-south direction, plan form, rough texture of wall surface, open space and court yards, landscape and vegetation all round to favourably reduce heat transfer and movements inside. But in the present era modern buildings, the features are RCC flat roofs with 3 m ceiling height, 10 cm wall thickness with smooth finishing, without an open space. In addition, the area is densely populated. Hence the inside room temperatures are relatively high due to heat transfer from conduction through walls and roof material, radiation through fenestration and from internal activities inside the building. Most of the modern buildings also have a glass façade which transmits intense heat loads. Comfortable living is based mainly on the inside room temperature and humidity levels and hence large amounts of energy is consumed in air-conditioning the buildings for a cool working and pleasant atmosphere. But energy consumption levels can be reduced by designing suitably for thermal and visual comfort by incorporating the passive techniques such as shading walls, roofs, fenestrations, provision of roof ponds, solar chimneys, trombe walls etc. Use of solar energy and rain water harvesting are also beneficial.

An attempt can be made to reduce the requirement of artificial illumination and air conditioning units by properly taking care in design of buildings such as orientation, length to breadth ratio, plan dimensions, wall material and thickness, opening areas, brick to glass surfaces, shading, projections etc.

A.7 DEVELOPMENT OF RURAL ENVIRONMENT

On July 14, 2004, President of India has laid plans for rural prosperity through connectivity. The following guidelines are noteworthy:

- Provide physical (roads, electronic and knowledge) connectivities which would result in economic activity.
- Provide facilities like sanitary, health, school education, self employment, and micro credit.
- Provide road network and transportation for connecting villages, electronic connectivity, health care services and education network for village cluster (20 to 40 villages) groups.

Give priorities for wasteland reclamation, herbal plantations, solar photo voltaic units, solid waste conversion to electricity, flood water and drought management, and defluoridation.

A.8 DRAFT NATIONAL ENVIRONMENT POLICY (AUGUST 2004)

(Released by the Union Ministry of Environment and Forests)

The salient features are:

- To bring in regulatory reforms
- To control environmental resources
- To set up environmental standards
- To adopt standardized management systems
- To promote environmental certification

The challenges are:

- Poverty
- Environmental degradation
- Global environmental concerns

The objectives are:

- Conservation of critical environmental resources
- Livelihood security for the poor
- Integration of environmental concerns in economic and social development
- Efficiency in using environmental resources
- Environmental governance
- Enhancement of resources for environmental conservation

The proposed regulations contain a set of regulatory institutions to work under a rigid legislative framework. Environmental concerns should be addressed at the planning stage of development projects to remove inadequacies later and to reduce costs. Further, accountability has to be fixed for non-compliance of regulations. No encroachment on forest lands is to be permitted. The coastal regulation zone notifications are to be reviewed to protect ecological systems. Heritage sites are to be protected while considering the development projects. Finally the environment is to be continuously monitored to enforce standards of environmental quality.

Some more field projects of interest are:

1. Water quality assessment at a source
2. Air quality monitoring near a traffic junction
3. Recycling of less foul liquid wastes in an institution
4. Sanitation measures in a colony

5. Demographic studies in a slum area.
6. Causes for disappearance of lakes.
7. Affects of tourism near the water bodies.
8. History of land site degradation.
9. Socio-economic changes due to the construction of a fly over in a town.

GLOSSARY

Abiotic	:	Non-living component in the environment
Abyssal zone	:	Cold, dark, bottom most layers in sea
Acclimatization	:	Adjustment to the changes in abiotic environmental conditions
Acid rain	:	Rain water turns acidic due to reaction with chemicals (SO_2, NO_2 or exhaust gases) in the moist atmosphere
Acute	:	Immediate affects due to exposure to some factors
Adaptation	:	Changes that allow organisms to live in a specified environment
Aerosol	:	Very fine particles or droplets in air
Albedo	:	Reflective property of a surface
Allergens	:	Substances that activate the immune system causing some reactions
Ambient atmosphere	:	The air component surrounding human environment
Antigens	:	Substances which react with some antibodies
Aquifer	:	Water yielding porous strata
Autotrophs	:	Which can synthesize food from in-organics (green plants) with an external energy source
BOD (Biochemical Oxygen Demand)	:	Amount of oxygen required by the organic matter under aerobic decomposition (mg/L) which may be a measure for strength of waste
Benthos	:	Living in deep waters
Biodegradation	:	Breaking down of organic material by microbial activity
Biogas	:	Methane produced in anaerobic decomposition of waste water

Biomagnification	:	Increase in concentration of some substances through food chains
Biomass	:	The total weight of all organisms in a given habitat i.e. the total material produced by living organisms
Biomes	:	Life zone with a specific climate, soil and other ecosystem characteristics restricting the types of organisms
Bio-remediation	:	Restoring environmental quality by using biological means
Biosphere	:	Surrounding atmosphere, lithosphere and hydrosphere supporting life
Biota	:	All organisms in a given area
COD (Chemical Oxygen Demand)	:	Amount of oxygen taken up for chemical oxidation by the waste water
Carrying capacity	:	The maximum potential that any ecosystem can support the living organisms
Cells	:	Living units making microbes, plants and animals
CFC (Chlorofluoro Carbon)	:	Releases chlorine which reacts with ozone in the atmosphere
Chronic	:	Long lasting results due to exposure of toxic substances
Climate	:	Weather conditions, consisting of annual changes in temperature, wind movements, rain fall etc. in a specified region
Coliform	:	Bacteria commonly found in the large intestines of human beings, indicates faecal contamination
Commensalism	:	Relationship in which some get benefits without harming others
Community	:	Natural populations living in the ecosystem of a given area
Composting	:	Biodegradation of organic matter to produce nutrient rich soil manure
Conservation	:	Resource management towards sustainability
Consumers	:	Organisms that obtain nutrients and energy from others
Coral reefs	:	Specific oceanic features found in tropical seas

Core	:	Huge molten hot mass (iron and nickel) at the centre of the earth
Crust	:	Outer shell of earth (even up to 50 km thickness) resting on mantle
DO (Dissolved Oxygen)	:	Amount of oxygen dissolved in a given volume of water at a given temperature and atmospheric pressure (mg/L)
Decomposers	:	Organisms which break complex organics into smaller molecules for their nutrient supply
Delta	:	Fertile land in a river basin
Demography	:	Study of populations, growth rate, age, structure, distribution and the related causes
Detritus	:	Fine matter containing decomposing solids.
Diversity	:	Presence of more organisms and more species, indicating richness of all groups
DNA	:	Deoxyribonucleic acid molecules causing the development of all cells
Ecological balance	:	The ability of nature to remain unchanged, readjusting with small disturbances
Ecology	:	Study of relationships between organisms and the environment
Ecosystem	:	A specific bio-system and its environmental conditions with interactions for living metabolism. A location where organisms live together with the environment
Ecotone	:	A transitional zone (boundary) between two ecological groups
Edaphic factors	:	They are the environmental conditions that determine the soil characteristics
Effluent	:	Waste water disposed into the atmosphere from a polluting industry
EIA	:	A study to report the environmental consequences due to implementation of a project
Emission standards	:	Permissible level of chemicals regulated for release into the atmosphere
Endangered species	:	Plants and animals in the danger of elimination completely
Endemic	:	Restricted to a specific region

Entropy	:	A measure of disorder which always try to increase
Environment	:	Surrounding conditions that affect the communities
Environmental development	:	Meeting today's needs and also preserving future needs
Environmental stability	:	Equilibrium conditions in an ecosystem
Environmental stress	:	Disturbances due to adverse environmental conditions
Epidemic	:	Spreading of diseases at a rapid rate in the environment
Epidemiology	:	Study of distribution of diseases and their causes and characteristics
Epilimnion	:	Warmer layers of waters
Erosion	:	Removal of material due to actions of water, wind etc
Estuary	:	Zone of river joining sea water
Euphotic zone	:	Upper layers of water bodies which supports photosynthesis
Eutrophic	:	Water rich in organic nutrients
Eutrophication	:	Rich in organic nutrients
Evolution	:	Process of gradual changes of species due to a number of factors
Exosphere	:	Outer layer beyond ionosphere
Fauna	:	Animals present in the region
Flora	:	Plants present in the region
Food chain	:	Sequence of transfer of food from one to another
Fossil fuels	:	Coal, petroleum and natural gas created under the ground over centuries
Fuel cell	:	Devices to produce electricity
Fugitive emissions	:	Substances such as dust entering into the atmosphere not from chimneys
Gene	:	Heredity representing unit cell
Geothermal energy	:	Energy obtained from the internal heat of earth's crust
Green revolution	:	Increasing agricultural production using more inputs
Greenhouse effect	:	Trapping of heat in the earth's atmosphere by CO_2, CFC and other gases to warm up the earth

Gross productivity	:	Sum total of all production in the nation
Habitat	:	A living place under given environmental conditions
Hazardous wastes	:	Toxic to life, ignitable, reactive
Heterotrophs	:	Organisms that feed on nutrients produced by other organisms
Homeostatic	:	Maintaining a sustainable steady state of living system
Humus	:	Brown residues after decomposition, contains soil nutrients
Hydrosphere	:	Water bodies like rivers, lakes, oceans
Hypolimnion	:	Cooler layers of water, unmixed
Igneous rocks	:	Granite and basalt like rocks formed due to solidification of molten magma (lava) in the interior earth
Indicator species	:	Which indicate the conditions in habitat
Infiltration	:	Percolation of water into soil
Insolation	:	Incoming solar radiation
Inversion	:	Temperature increases with height above the ground level
Ionosphere	:	Lower part of thermosphere
Irrigation	:	Supplying water to raise the crops
Landfill	:	A dumping site for the disposal of municipal garbage
Lentic	:	Standing water in lakes without movement
Limiting factors	:	Factors that regulate the growth and survival of living environment
Limnetic	:	Living in open water of a lake body
Lithosphere	:	Solid portion of earth including crust and mantle
Littoral	:	Shallow waters near shore
Logistic growth	:	Rate of increase in populations regulated by natural factors which establish equilibrium conditions based on resources
Mangroves	:	Salt tolerant forest lands like sunderbans
Mantle	:	Hot layers of rock in between outer crust and core of earth
Mesopause	:	Atmospheric layers in between mesosphere and thermosphere
Mesosphere	:	Middle layers of atmosphere with low temperatures

Metabolism	:	Life process
Metamorphic rocks	:	Modified igneous and sedimentary rocks due to physical and chemical reactions
Minerals	:	Naturally occurring inorganic solids with specific physical and chemical properties
Morbidity	:	Disease occurrence rate
Mortality	:	Death occurrence rate in populations
Mutation	:	Changes in the genetic cell material
Natural gas	:	Methane and hydrocarbons dissolved in petroleum products in oil bearing strata below earth's surface
Neurotoxins	:	Poisonous chemicals (Hg, Pb) to remove cells
Nuclear	:	Radioactive isotopes emitting radiation Fission – splitting into small atoms Fusion – two smaller atoms combine to release energy
Oligotrophic	:	Waters with less nutrients
Organic compounds	:	Containing carbon atoms in different arrangements
Oxygen sag	:	Oxygen levels in a river (distance – time) to indicate pollution levels at various stretches
Ozone	:	Highly reactive pollutant containing three oxygen atoms
PAN	:	Peroxyacetyl nitrate, a secondary pollutant formed due to reactions in the atmosphere
Parasites	:	Living on other organisms
Pathogens	:	Disease causing organisms
PCB	:	A toxic pollutant containing chlorinated hydrocarbons, biphenyl
pH	:	A measure of hydrogen ion concentration whereas acidity or alkalinity is a neutralization value
Photo voltaic cells	:	Energy conversion device to convert solar energy into electricity
Photosynthesis	:	A bio-chemical process in which green plants prepare food in the presence of CO_2, H_2O and chlorophyll
Planktons	:	Microscopic organism in the upper water layers

Pollutants	:	Harmful substances, injurious to life or material Primary – chemicals released directly from the source Secondary – harmful substances formed in atmosphere after chemical reactions
Pollution	:	To cause undesirable effects on life and materials
Predation	:	Feeding on others directly for survival
Producers	:	Organism that synthesizes food by itself from compounds (photosynthesis)
Profoundal	:	Deep, bottom layers
Putrefaction	:	Anaerobic decomposition of waste matter
Reclamation	:	Recovering contaminated sites into original form
Recycling	:	Processing waste products into useful ones
Red Data Book	:	Contains data on endangered species
Rock	:	A solid portion containing minerals
Run off	:	Water flow available on the land surface like rivers
Salinity	:	Amount of dissolved salts (NaCl) in water
Sanitation	:	Programmes to safeguard environmental health
Saprophytes	:	Plants living on dead organic matter
Sedimentary rocks	:	Compaction of sand and clay like material to form shale and sandstone
Sewage	:	Waste waters from domestic origin
Sludge	:	Solid matter in liquid effluents
Smog	:	A mixture of smoke and fog indicating photochemical oxidation involving NO_2, SO_2 and HC
Soil	:	Weathered surface material, unconsolidated
Solar energy	:	Extraction of energy from sun's radiation into heat or electricity
Species	:	Genetically similar organisms and their offspring
Stratopause	:	Atmospheric layers in between stratosphere and mesosphere
Stratosphere	:	Ozone rich layer with stable temperature about 25 km, from tropopause

Subsidence	:	Lowering of ground levels due to causes like more withdrawal of ground water
Succession	:	Series of changes in ecosystem over passage of time
Symbiosis	:	Living together of members of different species
Synergism	:	The sum of two effects is greater than their individual effects
Thermal pollution	:	Increase in water-body temperature which affects aquatic life
Thermocline	:	Temperature of transition zone between epilimnion and hypolimnion
Thermodynamics	:	Relationship between heat and work (matter and energy)
Thermosphere	:	Upper most part of atmosphere in which temperature increases with height (1500 km above ground surface)
Toxic	:	Poisonous chemicals very harmful even in dilute concentrations
Transpiration	:	Evaporation of water through plants
Trophic levels	:	The functional level or nutrient status of an organism in a food chain in an ecosystem
Tropopause	:	Atmospheric layers in between troposphere and stratosphere
Troposphere	:	Atmosphere near the earth (10 km) where temperature decreases with altitude
Turbidity	:	Obstruction to the passage of light in water due to fine colloidal suspensions
Urbanization	:	Circumstances leading to increasing concentration of population in towns and cities
Water logging	:	Saturation of soil with more water than required for crops
Water table	:	Upper level of water (zone of saturation) in the sub-surface soil
Watershed	:	Surface land area responsible for draining water into a river system
Weathering	:	Physical, mechanical or chemical changes in rocks due to wind, water, temperature or other reasons
Wetlands	:	Types of lands with standing water and vegetation

GENERAL QUESTIONS FOR STUDY AND ASSIGNMENT

1. Write a detailed note on the main components of biosphere and their utility for the welfare of mankind.
2. Explain briefly the biogeochemical aspects of carbon or Nitrogen cycle in the Nature.
3. List the advantages of conserving forest and wildlife resources.
4. Give a brief account of alternate resources to solve energy crisis.
5. Outline the steps to be taken to prevent depletion of ground water resources.
6. (a) Substantiate the statement 'land is a resource'.

   (b) Write a brief note on the mineral wealth of India.
7. What are the essential components of an ecosystem? Explain its structure and function.
8. How are food and energy needs satisfied in a pond ecosystem?
9. Terrestrial ecosystems are different at different places, why?
10. More than 70 per cent is salt water on the earth. What are the main functions of an ocean?
11. Justify the need for preservation of biodiversity in nature.
12. Mention the factors that threaten the biodiversity.
13. What are the steps taken by the government to conserve biodiversity?
14. What do you understand by the term demographic studies? List the demerits due to population explosion.
15. What is pollution? Name some pollutants and their harmful effects on human beings, plants and materials.
16. Enumerate the various measures taken in a community for the disposal of solid wastes.
17. Briefly outline the various methods adopted for the prevention and control of soil pollution.

18. What are the main characteristics of waste water? Mention the quality of water fit for drinking purposes. What are the measures taken for the prevention and control of water pollution?

19. Which are the sources that cause air pollution? List some primary and secondary pollutants and state their harmful effects on human beings.

20. Explain the term 'self purification property'. How is it helpful in preventing air pollution or river pollution?

21. State the ill effects due to depletion of ozone layer, and presence of green house gases in the atmosphere.

22. Industrialization, green revolution and urbanization have caused lot of pollution problems in the last fifty years? Explain and quantify the damage.

23. Explain the terms 'carrying capacity', and 'sustainability of the environment. Write a detailed note on the measures needed to maintain a sustainable environment.

24. 'Waste also is a resource'. Discuss the statement with suitable examples.

25. Describe the duties of a pollution control board to protect the environment as per the legal provisions.

26. Explain the terms 'human rights' and 'environmental ethics'.

27. If you are assigned a job of an 'Environmental Manager' in an industrial unit, how are you going to solve the various problems due to pollution?

28. Write notes on the interacting components of human environment in the biosphere.

29. State the issues related in the management of natural resources.

30. What is the index for measuring ecosystem biodiversity?

31. How are materials recycled in nature?

32. Mention the factors that affect the carrying capacity of the environment.

33. Is it possible to quantify the various aspects of sustainable development? Explain the assessment methods.

34. List out the possible effects of water pollution on freshwater ecosystems.

35. How are hazardous wastes disposed safely?

36. Write a note on the management aspects of minimizing waste production.

37. State the objectives of environmental quality standards and ISO 14000.

38. Briefly mention the steps taken by the government with regards to environmental protection.

39. Explain the need and scope for showing concerns to environmental problems.

40. Explain how overgrazing or mineral extraction leads to environmental degradation.

41. Write a note on the methods of conservation of natural resources.

42. Why are tropical rain forests more productive?

43. Explain the role played by oceans in regulating the environmental conditions.

44. Discuss the salient features of hot spots of biodiversity.

45. Explain the role of an individual and social organizations to prevent environmental pollution.

46. Specify the problems faced in disaster management.

47. Outline the various measures to attain sustainable environment.

48. Describe the practices of watershed management.

49. How are waste lands reclaimed?

50. Mention the environmental protection rules involved in implementing the act.